# Victory of the Earth

## Planet Battles and Development

By Steve Preston

2nd Edition

# Table of Contents

# What Do You Think You Know?

I'm tired of the scientific community blasting religion and the religious dogma blasting science. While those things will be important in this book, the most important element will be to RE-teach things that should have been presented in our schools or by scientific and religious leaders. You may think you have a pretty good handle on how the earth is the way it is, but after reading this book, hopefully, you will see that you have not been told the truth or at the very least, information you needed to make a reasonable attempt at the truth has not been provided to you. Let's first start by seeing what you know.

**Now with Nuclear decay timing being defunked, how can we determine when something happened?** You keep hearing about carbon 14 nuclear decay dating and about a dozen other nuclear decay dating systems and how accurate they are. They let us know exactly when dinosaurs roamed and planets evolved. Recent discoveries have halted all that. Sure some are trying desperately to hold on to them claiming they still are good or they are blinded by comfort, but there are other ways to find the truth.

**How did the Universe get here?-** You probably believe that there was a Big Bang and everything started or you believe that God said. "Let there be a universe!", and it started. Neither tells the whole story. There is a third more accurate answer.

**How did the Earth get in its position in the Universe?** You probably were told that Bode's law dictated that the earth should be in its position and so it was. That is so very wrong. Bode's "LAW???" should be struck from the textbooks. The "Log Law" should be put in its place and it also needs to be addressed with caution.

**How did animals get here?** You might believe that an amoeba mutated and a "survival of the fittest" evolution started everything to change until we are where we are. If not, you might believe that God made the animals a mere 6 thousand years ago or you believe that God created the animals over a very long time period. A 4th method makes more sense and combines science, physical evidence, ancient historical record, and religious rather than requiring one or the other to suffer.

**How did the Dinosaurs get so huge?** You possibly are thinking survival of the fittest again or God zapping the huge monsters into existence again. Let me make it perfectly clear, the Bible says no such thing and the survival of the fittest hypothesis would not have resulted in dinosaurs. While neither of these is the answer, there is a reasonable answer to this question.

**How did the huge mountain ranges develop on earth**? You probably were told and believe that plate tectonic action smashed pieces of the earth together and there they were. You will find that that answer is absurd. A much more reasonable answer awaits you in the book.

**How  and when did the Pacific Ocean get here**? You might believe that the earth was formed that way or believe that some massive meteor struck the earth. Neither appears to be correct. The correct answer has been well-studied and required one of the planets?

**How did Mars get split in half**? You might be saying, "If Mars had been torn in half, you would have heard about it before now." Unfortunately, the answer is you SHOULD HAVE been told about it.

**What is the asteroid belt?** You might think it was always in its current position, but, there is much more to the story.

**Are there remnants of buildings and cities on Mars?** If you say no, there is surprising evidence that may change your mind. If you say yes then when was the planet occupied and by whom?

**Did people live on the Moon and Venus during ancient times?** Like the Mars question, if you say no, there is surprising evidence that may change your mind. If you say yes then when were the places occupied and by whom?

**How and when did Venus turn into a ball of fire**?- If you believe that a thing we call the Greenhouse Effect destroyed Venus---You are wrong. If you believe that Venus caught on fire millions of years ago you are wrong again.

**When did the Huge gash that almost splits Venus in half occur**- You probably are thinking, "There is no such split!" While no one has told you about it, the huge split is important to earth's history.

**How did the Mammoths and other animals get quick frozen 10 thousand years ago**? If you believe an Ice Age did it, I'm sorry to say that is impossible. If you believe that mammoths lived in the arctic and just happened to freeze with temperate zone flowers in their mouths, you are wrong again. The answer is confirmed by many repeats of the same action and will be fully explained.

**Why are they finding many dinosaurs unfossilized?** Yes it means some dinosaurs were here no more than 20 thousand years ago, but it's not what you think. There was a dinosaur extinction well before that time.

**Were Neanderthals dumb**? If you believe that we are smarter than Neanderthal humans were, try to explain why their brains were larger. The explanation is simple.

**When did humans become civilized**? If you believe they became civilized with the beginning of the Bronze Age, that's not far enough back. If Neanderthals is your answer, you are still way off. If you believe that Adam was the first civilized human, again you are wrong and you have not read the Bible Genesis story completely. While the evidence is overwhelming and convincing, some will have a hard time with this major factor in earth's development.

**Was there a worldwide flood and did Noah carry 3 million animal types in his boat to safety**? I think I can explain the details of this time so that it will make sense and conform to historical record, Scientific discovery, and religion.

**What made people's life span decrease so dramatically 5 thousand years ago**? If you think that people always lived about 70 years, there is plenty of data that indicates much longer life spans. If you think it was the worldwide flood, you are wrong. By all evidence the flood occurred 10 thousand years ago and people lived long lives after that fateful occurrence. If you think people never lived longer than we do today, you are probably wrong again.

**When will global warming and the greenhouse catastrophe affect the earth**? If you believe soon, you are probably wrong again. If you believe the earth will not change in the near future you are wrong again. Prediction, scientific study, mathematical modeling, and physical evidence confirm the answer. With almost certainty, the earth will not get hot. The answer is scary and it is soon.

If you're looking for a Science book that ignores the more unpleasant things that have been witnessed either as physical evidence or by first hand descriptions and knowledge, you had better just shut the book right now. If you are looking for a book that will comfort you, you also have come to the wrong place. If you really want to learn about how the earth got to be the way it is today regardless of what you were told in the past, please continue. You will find that the lies and half-truths that you have been told don't change the past, they cover it up.

### Bizarre Is Not So Bizarre

Because people are not typically told about less than comfortable physical evidence, some things seem bizarre to us when we read about them. Hundreds of pieces of evidence tell us a different story than what you are used to reading. For instance-

*What would you think if I told you that Einstein's $E=MC^2$ formula is flawed?*

*What would you say if I told you that Dinosaurs weren't heavy?*

*What if I told you that Venus had a moon that affected the earth?*

*What if I told you the end of the Cretaceous Period was only 140 thousand years ago and there is substantial proof?*

*What if I told you Mars formed the Pacific Ocean?*

You would just think I'm crazy, but I'm not. The real problem is that you have not been told about how sketchy our definitions have been and how many anomalies have proven our old school descriptions were. This is the first step to opening your eyes and minds.

Below are a few of the seemingly inappropriate scientific assertions that will begin to make sense as you are presented the evidence that is typically ignored. They are a little different than the "theories" you have been taught. Rather than just accepting what you were taught, be prepared to be amazed at what is not normally told to us as we go through school. You have purposefully been kept in the dark about the world around you, but ignorance will not make truth go away.

*The Bible and science can and do together. I don't mean by twisting them around, but by using their innate truths to support one another.*

*Before the Big Bang there were at least 2 universes that did something scientists call the Big Splat. The 2 universes are still around according to modern scientists, mathematicians, and religious dogma. A single universe cannot exist on its own.*

*The Evolution Theory doesn't work. Nothing fits. The timing, the number of animals, the rapid expansion of animal life immediately following an extension period, the weird mistakes of nature, the lack or a transition model, the survival of the most unfit and many more curiosities all tell us the same thing. While it doesn't make sense, uncontrolled evolution of species is still taught as fact.*

*God creating all the animals doesn't make sense either and the Bible doesn't indicate that he did. We can even get an*

9

*approximation of what animals were created and those that came about another way. Those that were not "created" were deemed "abominations" or "unclean animals" in many ancient texts.*

*Like I previously stated, "E=MC²" doesn't work. Einstein knew and others know it today, but it is taught as if it were a law.*

*Invisible mass is a new concept for many, so I will explain what it is briefly and what it means to us. Mathematical string theorists require its existence and so does Religion.*

*Creationist theories don't work. While I don't go into this in detail, the disregarding of modern dating methodologies to make the claim that the world is only 6 thousand years old is not practical, reasonable, or founded. Oh yeah!—The theory is not Biblical either.*

*The continent of Pangea was not the super continent. There was more than one. The other massive continent has been named Prestonia but never discussed in school.*

*The Earth almost destroyed Mars. That incident made the Pacific Ocean. Mathematical models tell us this, physical evidence tells us this, and ancient historians told us this. Now I'm going to tell you and show you the details.*

*The Earth almost split in two several times. Huge Magma Mountains attest to the fact and the danger. Here's the kicker. The splits have occurred on the opposite side of the world as huge craters that are left by gigantic meteors.*

*The Earth has flipped on is rotational axis many times. When it flips, it does it quickly and animals are quickly frozen or twisted together in huge piles. Surely you have heard about the frozen animals, but have you been told how it was possible to "quick freeze" a massive animal like a Mammoth?*

*Civilized humans lived with the dinosaurs. That doesn't mean that dinosaurs were living recently, it means that humans have been here a long time and the proof is amazing. ---No! The humans did not come from outer space. If we look at ancient books, we can find the truth that agrees with huge quantities of physical evidence about this fact.*

*The moon of Venus exploded 10 thousand years ago. Yes!- I said its moon. When the explosion occurred, Venus caught on fire and earth was pelted with thousands and thousands of meteorites. Venus itself almost split in two and the evidence is everywhere. We can even date the event and we can get agreement in ancient religious texts around the world including Jewish texts. All this happened about 10 thousand years ago.*

*The world is not heading for a meltdown like Venus. Instead, a terrible Ice Age will be here within our lifetime. Scientists have been recording details of this event now for years.*

## Absurdness Recognition

I know all these things sound absurd right now, but soon I think you will begin to see what has been "hidden" from you. With that being said, Let's get started. What better place to start than 15 Billion years ago. It was the beginning of the beginning so to speak.

**Section 1**-The first part of this book will deal with Macrocosmic definitions being used today to support a truer description of how things got here.

**Section 2**-The next section will be a description of the beginnings of our solar system and the earth.

**Section 3**- Following the overview of earth problems, I will concentrate on the affect Mars had on the Earth.

**Section 4**-After the earth and Mars settle down, there is a discussion of the key elements that molded the earth. Meteors, rotational anomalies, volcanoes, and the earth splitting all characterized the major portion of earth's history. Sometimes their details are ignored to make us feel safe.

**Section 6**-Evolution is tackled next with surprising evidence and conclusions.

**Section 7**-The evidence of an ancient civilized human existence on the earth follows the evolution discussions.

**Section 8**-The effect of the Venusian Moon exploding is the next topic. The evidence will certainly surprise you.

**Section 9-**The Worldwide Flood identified in the Bible is discussed next. The evidence may open your eyes and bring your religious beliefs you're your scientific understanding closer together.

**Section 10-**How people got electricity 7 thousand years ago is the next topic. Scientists have known for years that people had an electrical source. This is a possible answer to the problem of HOW and what did they do with it.

**Section 11-**The first world war of the Holocene Age is the next to last topic, but it is very important as we try to determine why people only use 10 percent of their brains. I know it doesn't seem like a tower has anything to do with the brain, but you will again be surprised.

**Section 12-**The last section is on the upcoming ice age. –Get out your winter clothes.

If you are you ready, let's look at the Big Picture.

# Solar System Timeline

I know that you have been told that the Solar system and the planets that make it up were put into place some 3 to 5 billion years ago and generally were and are still positioned in accordance with Bode's Law. Unfortunately, the evidence does not support these elements. In order to establish this view of the Solar system, many pieces of evidence had to be ignored. This was all done in order to keep you at peace and make our neighboring planets less complicated. Hopefully you are not interested in this less complicated view. This overview does not answer all the questions, but it; at least it presents a forum from which questions can logically be addressed.

The timeline following will seem wrong to you, because you have not been given all of the facts. I am not trying to indicate that this book contains all the RIGHT assumptions, but it does not throw away evidence to force a history that is riddled with inconsistencies, half-truths, and even lies.

## Years Ago-Chain of Events

**-4 Billion?**-The Universe made a big bang of some kind. **[Pros and cons of this theory are presented, but the timeframe of the establishment of the universe is generally accepted around this time.]**

**-4.5 million** -The planet positions were unstable. Jupiter almost ignited to become a star. Saturn, Neptune and Uranus were spinning so fast that pieces of their respective surface masses were thrown into spinning clouds of debris that rotated around them. The Earth "super planet" was one of these.

**-400 thousand-** [**End of the Permian Age**] Mars and Earth had a close encounter and the gravitation pressure on both planets formed mountains. [**The "Plate Tectonic" version of this event doesn't work.**]

**-350 thousand-**Mars' and Earth's revolution paths brought them close to one another again and mountains were pulled upward again on both planets [**Math models of this and subsequent events confirm this probable action.**]

**-400 thousand- End of the Triassic Age**] Mars and Earth had still another close encounter. This time both planets were ripped apart. Much of the life on both planets died. While earth had a huge chunk pulled away, Mars was ripped in half. Earth began to repair itself. The last remaining landmass, Pangea, broke apart and slowly shifted to the huge hole left by the Mars encounter.

**-300 thousand-**Ancient Humans were living on earth by this time. [**An undeniable list of physical evidence makes this theory the only logical choice**]

**-220 thousand-** [**End of the Jurassic Age**]Some catastrophe occurred and the Earth began slowing down on its axis. Everything got heavier as gravity increased. Large animals began to have a difficult time.   [**Physical evidence supports this theory**].

**-160 thousand-** Perhaps because of the slowdown, people became afraid. Evidence suggests that people from earth set up colonies on Mars and other planets. [**Not little green men as has been comically referred**]

**-140 thousand-**The ancient humans evolved and became strange beings called angels. [**While you have been told that angels were always here, there is ample proof of another probability.**]

**-120 thousand-**[**End of the Cretaceous Age**] A large quantity of very large meteorites hit the earth. When they hit, the earth split open and tons of magma were expelled. This action formed most of what is now the country of India. This is the last straw for the Large dinosaurs. During this same time, a huge war was waged between 2 major groups. One was made up of many of the new "Angels" and the others living in a place called heaven. The heaven side won and the losers were turned back into humans and had to return to earth to live.

**-110 thousand-**Outposts on the Moon, Venus and Mars along with other possible sites were reestablished

**-100 thousand-**The atmosphere of Mars slowly dissipated along with its water. The inhabitants died or left.

**-90 thousand-**The new humans begin making new animal types and begin experimenting with apes. Part of the experimentation involved transferring semen and human eggs between the apes and humans. The ape-man was made. The last of these was called the homo-habilis ape-man.

**80 thousand-**God created a new human that was somewhat similar to the ape-man, but substantially different. This was the homo-erectus man. As it normally does, the Earth flipped on its axis and caused substantial and immediate climate changes and a lot of tsunamis, volcanoes and storms. [**Physical evidence shows that this is a fairly normal occurrence.**]

**70 thousand-**The used to be angel humans had sex with the homo erectus to produce a hybrid called Neanderthalis.

**-40 thousand-** God again steps in and created a new type of human this time he made one that had a soul that could live in heaven. This was the Adamic man or Cro-Magnon.

**-15 thousand-**A worldwide war erupted. Millions of people died.

**-11Thousand-**Something happened to affect Venus. Possibly it was involved in the still ongoing war. The Venusian moon shattered. Meteorites from the explosion hit along the equatorial section of Earth at that time. [**Substantial physical evidence supports this theory.**]-Venus turned into a fireball. Very few of its inhabitants survived, but a few reached the earth. Venus may have even been pushed a little closer to the sun during this terrible event. Earth wasn't left out of the catastrophe. It became somewhat unstable and changed its rotation again. Water levels rose as the polar icecaps were repositioned and the Island nations

15

were submerged. **[Huge amounts of written and physical evidence supports this theory.]**

**-10 Thousand- [End of the Pleistocene Age]**Different from the event of 11 thousand years ago, a worldwide flood comes. Most likely, it was the result of a comet strike and another earth axis shift happening about the same time. A few humans learned of the upcoming event and were saved. Most perished. **[Huge amounts of written and physical evidence supports this theory.]**

**-5.5 Thousand-**A battle station and tower were built in a place called Shinar. Now called the Tower of Babel, another huge war was waged for many years. The war seems to be centered on the tower. Some of the inhabitants were stationed on the moon during this time. Some type of accident occurred and humans lost many of their capabilities from the spreading virus. Only the Lunar colonists were unaffected. Even the brain size of the inhabitants began to shrink. **[Huge amounts of written and physical evidence supports this theory.]**

**Today-**Man struggled to regain civilization. He relearned about genetic manipulation. He relearned about nuclear weapons. He even began to venture on to nearby planets.

**+25 years in the future** Within the next 25 years a large meteor is predicted to hit earth. The prediction is from Mother Shipton, Nostradamus, the Bible and many others. This begins a new "Dark Age" and "Ice Age" on earth.

# 4.5 Billion Years Ago

So you think a book should start from the beginning so I will try one of the beginnings. Just about everyone has already determined what he believes to be truth. With respect to our beginnings, some believe the Big Bang Theory. Some believe the "interpretively literal" "Creation Story" in 7 days. Some believe Survival of the Fittest and Darwin's Evolution Theory. Some believe everything their government tells them. Some believe everything their clergy tells them. Some believe what they feel in their heart. Some believe only what they see. Some people don't believe in anything. It is time to open our eyes and look at evidence to support or not support our "TRUTH".

## Here is a Little Confusion

Einstein was asked, *"If a tree falls in the forest and no one was around to hear it, did it make a sound?"* His answer may amaze some as he said, *"There is no tree!"* Relativisticly, the tree would not be in the reality of an observer not going the same velocity. Vibrationally, mass has no real meaning, instead it is simply vibrational resonance of Aether that has no mass. Anthropicly, the tree is not needed in a world without cognizant appreciation, so it would not exist. These three major sciences were augmented by Quantum Physics which tears apart time and space as constants so that only positional uncertainty is left. Man o-man! This is way confusing. If these smart theoretical Physicists can't help us we had better go it alone.

Our gut tells us there was a time when our 2 universes came together and forced a massive exchange in energy. Big Splat/Big Bang; whatever happened, 90 % of the universes were formed within a very short time. If you thought we only had a single universe, let me tell you that no conceptual theorist of this century has a model without at least 2 universes. The 2 universe model is the simplest so that is what I need to use. One universe goes forward in time and has what we call mass while the "Super Symmetric" universe is going backwards in time to us and is filled with "anti-matter. It has to be that was for all the math models to work. For years Albert Einstein was in a state of despair believing he had no idea about the universe. He claimed that if we send out photons of light , they will travel to the end of the universe and go into nothingness. After a time, there would be no more LIGHT in the entire universe. The same could be said for time. If time travels to the end and disappears, soon we would have not time and mass would have that same fate.

## Eureka!

Dr. Milo Wolffe would clear the air when viewing a locked together universe that was completely opposite. As energy left our universe it would enter the other one. Time would be a continuous backwards and forwards loop, and Mass would turn into energy as it departed our universe and Mass leaving our adjacent universe would become Energy for us.

I know that has not gotten our story going, but I, at least wanted you to understand how important the universe of Heaven is to our existence. Here is an interesting point. At the end of the battle of Armageddon, the entire Earth is destroyed, but the Bible says, God had to make a new HEAVEN and a new EARTH. There had not been a battle in Heaven, but is got destroyed by "super symmetry" of our universes.

### *Red Shift Dilemma*

Video spectrum red shifts of quasars are too great. If you didn't know about red-shift, Let me give you a quick definition. If something is going away from you at near the speed of light, Einstein's law of Relativity indicated that everything emitted from that object would have its wavelength lowered [shifted red]. By the way if you wondered why the Hubble telescope is named that, it is for Dr. Hubble who was the first to prove this on stars racing away from us.

These red shifts indicate that we are able to view stars that are almost 15 billion years old and at unbelievable distances away when using the big bang model. The stars would have to have been larger, brighter, and more dense that anything that we could possibly imagine. The problem is that **several** of these supposedly immense stars have been located. They all have the same anomaly and everyone is unbelievable, according to "Time-Life", Stars, 1988. To make this more impossible, all stars we see are red shifted the same all the way around us showing we ARE the CENTER OF THE BIG BANG. If the earth existed then, it would have been evaporated. That is where Anthropic science comes in, but don't worry about that so we can get going.

---

*As a model, one can believe that the Big Bang actually did occur and it is a great way to describe what happened, but there must have been some outside force that helped it happen and shaped its outcome. There also must be answers to the questions above.*

---

Before we go on, let's have a little fun with Einstein.

## Energy Laws

To make this complementary nature seem more real let's look at the elementary physical element equations. They all act identically, because they all represent the same thing.

$$E= \tfrac{1}{2} KX^2 \text{ [universal law of potential energy]}$$

$$E= \tfrac{1}{2} LI^2 \text{ [universal law of magnetic energy]}$$

$$E= \tfrac{1}{2} CV^2 \text{ [universal law of capacitive energy]}$$

$$\underline{E= \tfrac{1}{2} MC^2 \text{ [possible universal law of mass energy]}}$$

I know that the Einstein version of the mass equation is different in the above equations [the ½ is missing], but that might make sense if the energy associated with mass was shared in two universes as is now being recognized.

## M-Theory

By the way, there is another thing we should look at just a little and that is what scientists call the M-Theory. [Membrane Theory] If you could view this it would show that between each universe there is a "membrane to separate characteristics. To hold these characteristics, it was modeled that they were each made up of 11 dimensions rather than the only 3 dimensions we were taught in school. Mass had 3 dimensions, the energy coming into our universe as mass left our adjacent one had its own 3 dimensions, consciousness and life was described by 3 dimensions and time was characterized by the duality I mentioned before.

## Planets and an apology

Hopefully, you have had enough of this background, because I'm having enough trouble with 2 universes. I thought that it was important to establish the basis of all particles so that we can better understand the basis for planetary makeup. I wanted to make you understand that scientists that tell us what they think we want to hear concerning history and everything else. That doesn't mean that what we have been told has any merit. If it did, there would not be so many anomalies in the physical evidence. The same is true for history and religious dogma. I'm sorry for the not warning you about the science stuff, but I didn't know about it until it was too late.

## How Old is the Earth Anyway?

Some may say the world and the rest of the planets in our solar system are only 6 thousands years old, but the evidence does not support it. A more probable age for the planets originally was determined to be about 4.55 billion years, which goes along with the estimated time frame of the big-bang theory previously discussed. Unfortunately, this 4.5 billion year value is derived f by looking at radioactive lead isotopes. Lead was believed to be the most useful. Not only did the material exhibit half-lives that were extremely long, but also because there were 4 different, easily found, variants including Pb-206, Pb-207, Pb-208 and Pb-204; so you could do cross-comparison of data within the sample itself. Besides the lead testing, the ages of meteorites have been tested by a wide assortment of radioactive dating materials. Of those, "Rb-Sr" is the most effective and most used isotope comparison-testing group. Anyway, the age of the Earth at 4.5 billion years is now under attack as radio activity standards have been destroyed. Best we can guess is that the initial estimates are off by a factor of a thousand so the earth is probably **more on the line of 4.5 million years old**. I'll explain the confusion soon, but right now understand that the Earth got here a long time ago.

Please understand; relying on scientific theory "only" is not a responsible way to gain truth. Let me give you an example.

# Scientific Conclusion Example

Here is an example of how science works and how scientists can be EASILY mislead. This example centers on true statements about "eyes". When investigating the number of eyes that a spider had, Scientists discovered that it had eight eyes to match the number of legs that it had. Again the investigation continued. *"How many eyes does a starfish have?"*, was questioned. It was discovered that the starfish had 5 legs and 5 corresponding eyes. Humans were investigated next with common results two legs and two eyes. The emboldened scientists proposed the *"Leg to Eye Count Theory"* which stated that the *"quantity of legs is proportional to eye quantity"*. They finally tested the theory against the bee, but everything fell apart. *"Six legs must mean 6 eyes"*, they insisted. Well, the bee actually has 5 eyes, 2 regular eyes and 3 light sensing eyes on top of its head. To make the theory work, all the scientists had to do was cut off one of their legs. [Just like anomalous evidence was removed or simply not examined in order to build the theory of evolution.]

*For many scientists, the statement might be said---"If a theory doesn't work; cut off one of the legs."*

I'll try not to cut off any legs here. Of course there is no "eye/leg" count theory, but it illustrates that, many times, science is quick to use limited testing to prove a theory and will do just about anything to modify facts so that they conform with pre-established doctrine. By the way, I have no idea why starfish and bees have 5 eyes.

## Out of Place Objects

The reason I brought up the example above is that scientists have continually misrepresented evidence when it did not FIT their theories.

The topic of this work is the geologic and catastrophic development of the various planets that make up our solar system. Before we can completely understand some of the otherwise anomalous evidence that has been found, and many times has been discarded, we must bring up something that will not sit well with many. While many anomalous bits of evidence have been destroyed for the cause of scientific comfort, every year more anomalous things are found. These things point to one thing.

### Civilized Man Was On Earth Eons Ago

This concept is important as we go forward investigation how the planets became what they are today. The people on earth occasionally affected their development or artifacts left by these individuals help us verify elements of the various planetary developments. Below is a short list of some of the evidence that has been found and proves that civilized man was on earth during a very ancient time.

- *Modern human bones inside coal and rock*

- *Man's footprints and shoeprints with dinosaur tracks,*

- *Cups, gold chains, nails, mortars, and hammers found in bocks of coal or inside stones,*

- *Batteries, Gold rings, and other manmade objects found inside geodes*

- *Walls, Flooring, and strange writing, found deep in the Earth,*

- *And many other artifacts*

There is a short section following to provide a little more information about some of the finds. These anomalies are found around the world. Besides in this book, their details can be easily found on the Internet, in various museums, and in a few historical books. The concept of ancient human habitation of the earth is

NEVER presented in our schools and therefore, most of the discoveries today must be shadowed with hyperbole and grandiose theories which hide a more realistic probability.

## Land Bridge Theory

One comical theory is that People from Europe traveled through Asia and across an Ice-land bridge connecting Asia and North America. Once they were on the Bridge, they air ice and kept with fire they made from sticks [wait a minute-no wood] I mean, they made do by huddling together in a great bunch until they walked for a hundred years or so to get to Ohio. I'm sorry!!! that never happened.

## Plate Tectonic Mountain Theory

Another one that is sort of comical has to do with plate tectonics. Certainly plate tectonics moves the surface of our planet around and makes volcanoes, but someone got the idea that the entire mountain range from the Antarctica, through South America, Through North America, And down along the Asia border was made as a massive plate pushed billions of miles of earth upward. Here is some problems. The first is dynamic what type of action would initiate this thing?---Nothing on earth has the energy. The second one is in direction. Let's say some volcano the size of the united States erupted in--- and we have a problem, The plate can only go in one direction and the mountain range goes well over ½ way around the earth. The plate would be going in 2 directions at the same time.

## The Old Mammoth Story

Let me give you one more ridiculous theory concerning massive mammoths found quick frozen in Siberia with flowers still in their mouth. A popular theory is that Elephants lied Siberia and Flowers used to grow their before they realized that cold weather was not the thing for flowers. The Mammoths simply stood still long enough to finally die and they kept flowers in their mouth for later. Instead of rotting they were quick frozen as ----- I never understood that part but there are tens of thousands of these mammoths quick frozen for some good reason.

## There is a new one going around

Recently, all types of remains from huge dinosaurs have been showing up and they are not fossilized. To make it more strange, many are so radioactive that the bones must be painted with lead paints to reduce the emissions after many years have passed since they died. Actually, I have not heard a silly theory about these yet as the finds are still only a few years old but I'll bet it will be a doosey. I'm going to give you "reasonable" alternatives to these as we look for actual truth rather than the vain truths developed for those wanting a comfortable answer and will allow for silly.

Speaking of sill have you ever been told light is sometimes a wave and sometimes a particle? If you had any college science, that is still what they say. I know you want to laugh, but be kind and I'll try to help[with that one as well. First let's retime the earth.

# Ice Core Dating

As I develop this history we will find that people have been in North America for a long time and even walked with dinosaurs. That being said, I'm not talking about a 100 million years, I'm talking about 150 thousand or so. All of the basic timing of the Earth was in jeopardy. This is even more pronounced in the past as the solar actions would vary drastically as the distances shifted and atmospherics changed. Nuclear decay dating told us one thing, but other dating methods would not concur. Today we know that the nuclear decay dating of things including Electron Spin Dating and Uranium Dating, Thorium Protactinium Dating, Oxygen Sediment Dating, Lead-lead-lead Dating, and Argon Dating [which we originally used to date the ages of the Earth] are flawed. The old standard carbon 14 dating also seemed in jeopardy. Dating beyond about 30 thousand years was most likely much younger than tested. If there had been nuclear events [bombs or even volcanic eruptions] the timing was also changed drastically. Other methods had to be employed to determine how everything should be timed, but classroom information was not changed. That would confuse the students.

## Hawaii Hotspot Track

Hawaii is not a tiny group of islands, but instead is an indicator of where the Earth magma has a hotspot. As the crust moves differently than the stuff below, the hotspot relative to the crust moves and each time the hotspot burns through another piece of crust, a volcano erupts which seals off the area after a time and an island is made for a few thousand years. This travelling hotspot known as Hawaii is show next. The descriptions provided shows what was happening along the way. Because the hotspot moves perpendicular to the axis of the Earth we also know how the earth was spinning as shown by the arrows, but the actual timing is not described here.

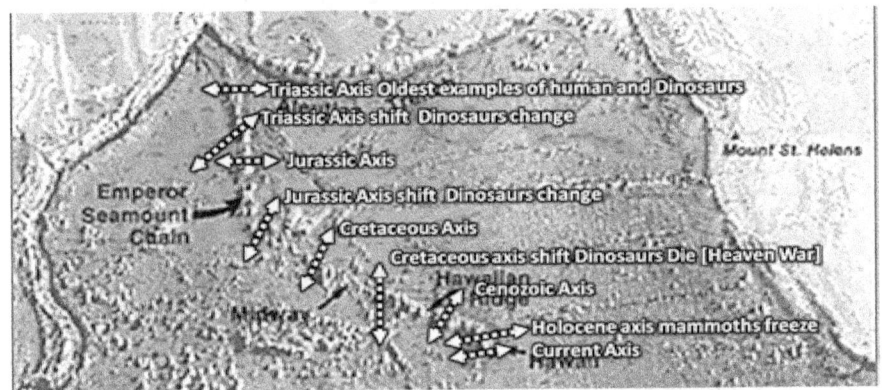

Luckily, we have a good timing agent anywhere there is a large accumulation of Ice. In Greenland and Antarctica, researchers bore into the ice and look for the fine changes that represent seasons to determine how cold or how much $CO_2$ or whatever. These tests all show the same thing as indicated in the next chart. Every 120 thousand years or so, there is a marked change in temperature, $CO_2$, dust, and deuterium etc.

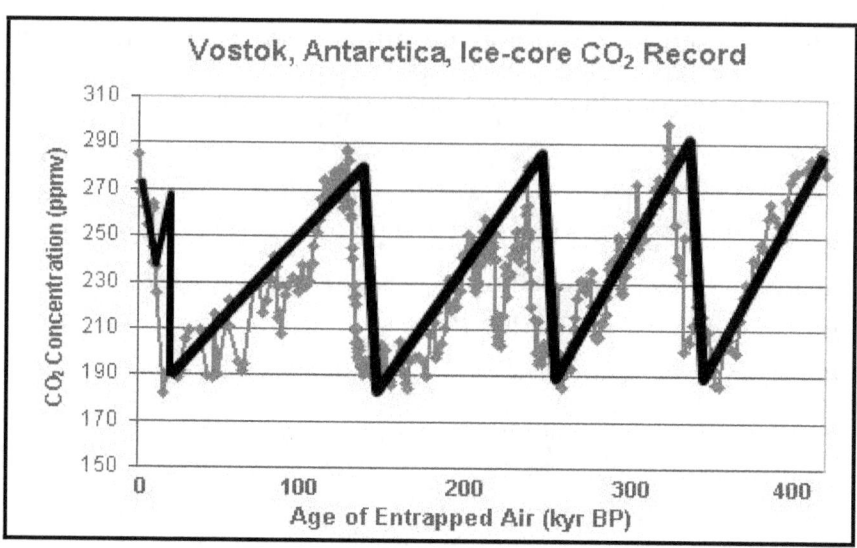

The changes match up to earth axis shifts. Every time the Earth shifts, something bad happens. Animals become extinct, massive floods cover the world etc. If we compare the timing of the last shift on the hotspot trail, knowing that was only 10 thousand years ago when the Mammoths froze instantly as they were shifted to the Arctic we can make a cursory determination of timing knowing the trails have moved 100 miles in the last 10 thousand years. The timing is shown below.

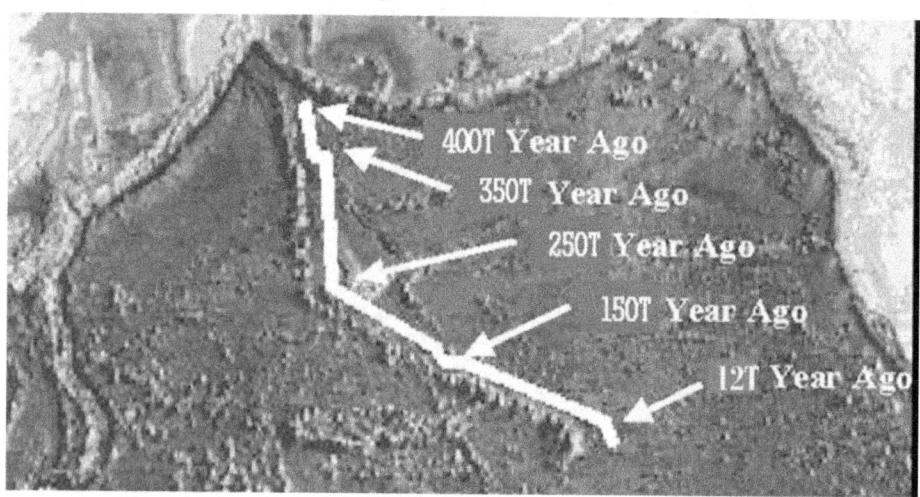

Certainly, this is no way to confirm timing, so let's see how the "shift times" match up to the Ice cores. Using the shift timing noted above, we can test the time against Ice Cores as shown next.

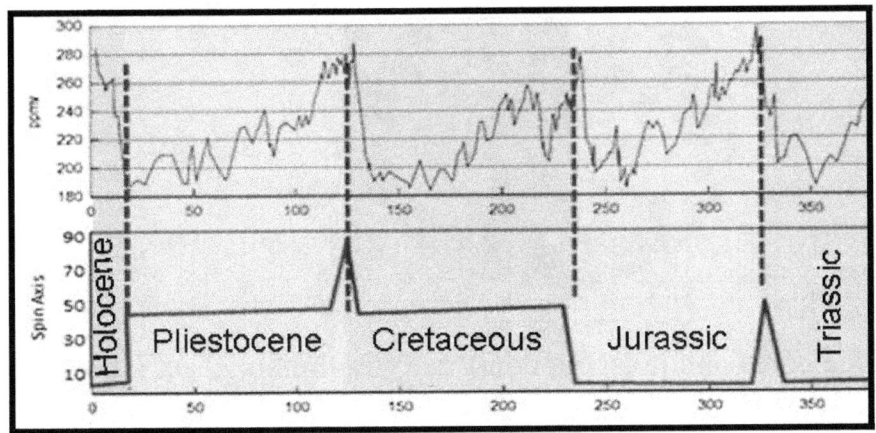

By this, the Dinosaurs did not die 65 million years ago, but instead, the event called the K-T layer [K stands for iridium chalk] happened only 100 to 150 thousand years ago rather than 65 Million. I know that is a big difference, but it is the best information we have to date.

Notice that for a few thousand years about 100 thousand years ago Antarctica was probably warm between the Jurassic and Cretaceous Period. Sure enough animals from that time have been found under the ice---just sitting there waiting to be found. The graphic below tries to show some possible major earth "settling" points and general information about those spin axes. For instance, notice that the earth spin goes along the east coast of the United States 10 thousand years ago. This will be important later as we piece all of this together to try to see critical time marks to help us reevaluate the time line for us as humans.

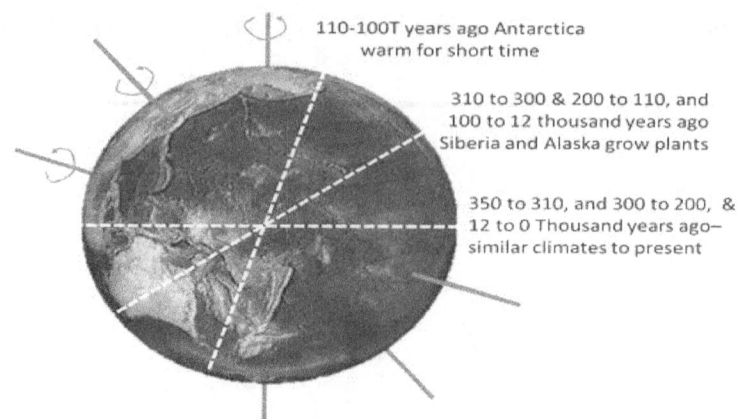

110-100T years ago Antarctica warm for short time

310 to 300 & 200 to 110, and 100 to 12 thousand years ago Siberia and Alaska grow plants

350 to 310, and 300 to 200, & 12 to 0 Thousand years ago— similar climates to present

If Ice cores and hotspots don't get you thinking, let's check out shells.

## Marine Isotope Stage [MIS] Timing

Some people may still be reluctant to give up what the schools have been preaching so very long, so I thought I would bring out one last attempt at presenting sanity. Large numbers of scientists around the globe are doing Marine Isotope Stage timing by digging in dirt. It seems looking at the levels of Oxygen 18 shows how hot or cold a point is in time while checking relative Oxygen 18 isotopes in Calcite [which just happens to be the main ingredient in seashells], one can tell just how many of the things were here during each period. Checking around the globe has given us a good map about climate and number of seashell, which correlates to number of animals in general so it is easy to see where extinction periods are. Guess where they line up? Time's up! They are an almost exact match as shown below. MIS levels are shown next above the ice core sample from Antarctica. Massive drops in $O_{18}$ mean massive drops in sea shells and all other life. Notice there is no extinction period between the Tertiary and Pleistocene Ages marked by Cro-Magnon appearing.

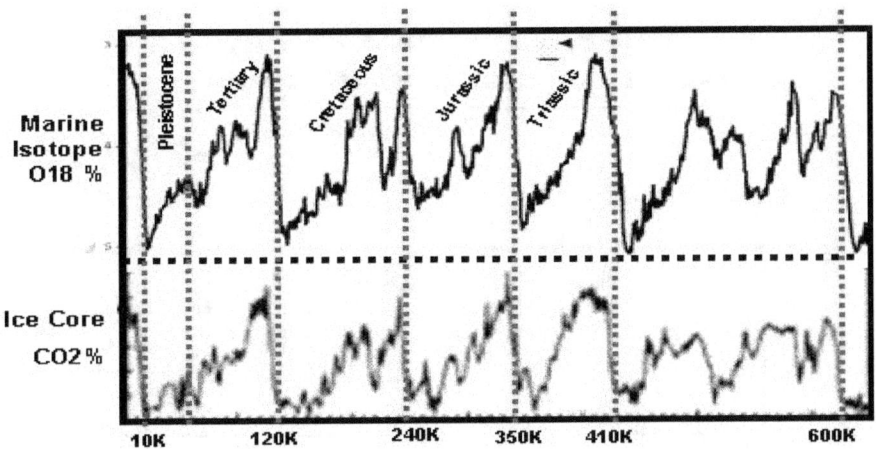

Please tell me you see some similarity in these charts!!!!!- so I can get to explaining other things faster. Before we leave this chart, please notice a sharp rise about 11 thousand years ago followed by a dip around 10 thousand years ago which indicates 2 massive climate changes occurred within a relatively short time. Yes, I know there is one of these things 220 thousand years ago, but for this history, the more recent ones are of more importance.

## Greenland Check

From the next chart, we can see a correlation in near term events. 11 thousand years ago a major spike in temperature with a fast cooling followed by another just a thousand years later then an almost flat plateau where Greenland's temperature has not changed and Greenland's position relative to the axis of spin has been unchanged. Before that time, it seems, the temperature was generally colder with what looks like a rise in temperature starting around 100 thousand years ago.

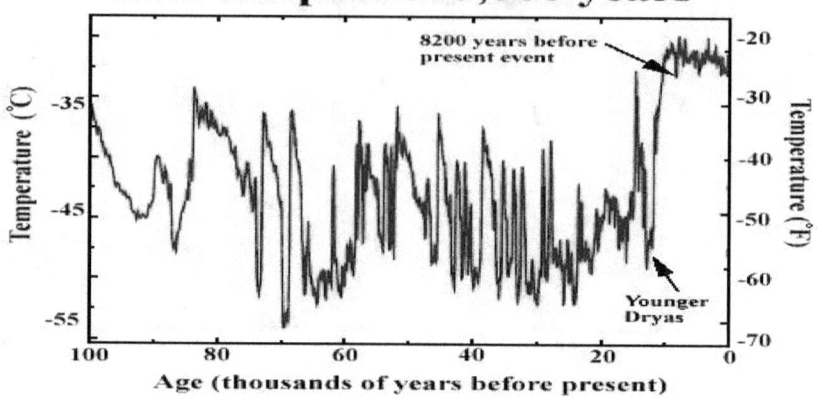

## Temperatures in Greenland over the past 100,000 years

8200 years before present event

Younger Dryas

Temperature (°C)

Temperature (°F)

Age (thousands of years before present)

## Paleo-Magnetic Confirmation

That brings us to PaleoMagnetic Dating. The Atlantic Ocean is getting wider about an inch a year, averaged worldwide. While the building of the great mountains has little to do with the normal tectonic plate "drift", we can pretty accurately measure the widening ocean in various ways including measuring distances between matched magnetic landmarks on either side of a widening gap on the ocean floor. The Old theory indicated that 180 million years ago the continent Pangea began splitting apart and has been drifting ever since. In so doing, the landmasses of the Western and Eastern hemispheres separated and opened the Atlantic Ocean basin today.

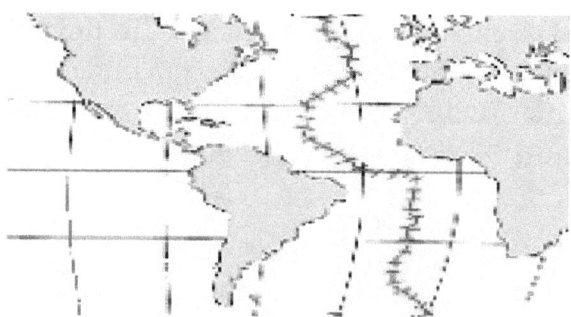

Plate tectonics tells us the outer hard crust of Earth consists actually of a dozen or so distinct, hard plates that drift individually on hot, deformable rock. An unequal distribution of heat within Earth moves the plates. The boundary between the

32

plates forming the Atlantic Ocean is smack down the middle along the Mid-Atlantic Ridge, shown as the hashed line in the figure above. The ridge is where we must look to find a widening gap, which accounts for the widening ocean. That is where we measure the rate of separation.

Where the plates separate, white-hot soft mantle oozes up from great depths within the Earth to fill the gap. The molten rock cools slowly into new slivers of sea floor. This happened over and over again through the eons. That's how the Atlantic Ocean widened-by a spreading sea floor. We measure the gap rate in various ways including direct measurements of plate movement using satellite images. Another is the Paleo-magnetic method. As the Earth's magnetic poles reverse polarity periodically, the North Pole becomes the South Pole and vice versa and much of the magma spewing out is iron.

Iron-rich rock has a peculiar property: if you heat it above its curie point of 580 degrees Centigrade, it loses its magnetism. When it cools, the rock gets re-magnetized in the direction of the existing Earth's magnetic field. So it's a magnet with the poles aligning with the poles of the Earth at the time of the cooling. The neat thing about this is; the magnetic field of the rock, once cooled, stays frozen in this orientation. It becomes a record of the Earth's field at the time of its cooling.

To measure the rate of separation, we identify two slivers of sea floor on opposite sides of the ridge that have the same magnetic polarities frozen at the same time. If you know when these reversals occur, one can simply measure the distance between magnetic alignments of the ocean floor and one can determine the rate of expansion and how long ago Pangea began to separate. Unfortunately, if the initial time-base is wrong everything is skewed. With that, let's look at the center of the Atlantic Ocean. The graph following shows the last 14 flips in relative position over "an unknown time", but when we match it up with ice, something magical occurs.

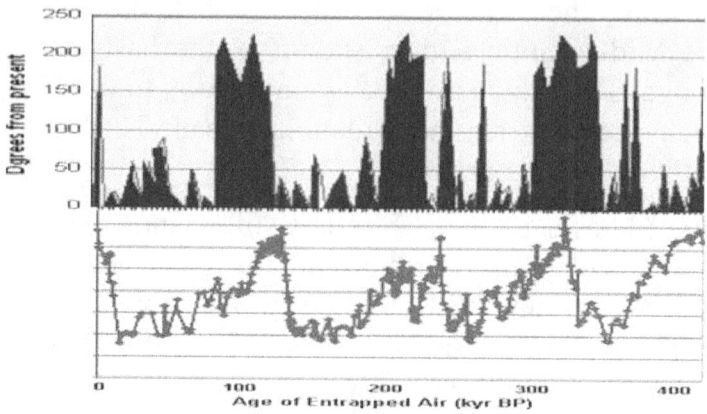

Changes in the earth axis correlate very well with the data from the Ice core testing when the data is compressed. I know I haven't given you a real good reason to compress the data, but you certainly should recognize that the old data was substantially unreliable. Here is what you should recognize. The magnetic field reversals and the cyclic ice core $CO_2$ levels seem to have a repetitive, cyclic nature. Even that strange change around 230 thousand years ago seems to correlate with the mid Atlantic data. I need you to notice one more thing. The compressed timing gives us more substantiation for 2 major climactic events occurring within only a very short period of time around 10 thousand and 11 thousand years ago. These 2 events are well known. The last one is known as the end of the Pleistocene Age as the Earth shifted drastically to quick freeze Mammoths in Siberia such that they still had flowers in their mouths that could never survive in Siberia. The event that happened a thousand years earlier may be even more important to our discussion in North America as hundreds of thousands of meteors struck the East coast and changed America forever.

## Some Don't Believe in Shifting Poles

Some people try to infer that this whole thing about the Earth changing its axis is hogwash. Well, I think that there is just way too much data to assume otherwise. Antarctica, with its dinosaur bones, the quick frozen Mammoths, the various polarities of the deposited iron from volcanic action in the middle of the Atlantic Ocean; they all tell the same story. The Earth axis can move and with it, there can be relatively fast and devastating climatic changes. These changes are horrible, and usually are the most powerful agent for extinction. If we are looking for a cause for the Earth axis shifting is from Comet or Meteor bombardment. Whenever a comet or meteors hit and the earth axis shifts right afterwards, total chaos occurs as it did about 120 thousand years ago then 11 thousand years ago followed by another possible attack 10 thousand years ago. These 3 dates are important to us as humans along with a deadly time 6 thousand years ago that was not impacted by the shifting of the Earth. A hundred thousand years ago makes the extinction of most of the dinosaurs and most of the human race at that time. After the extinction, the Bible indicated that the earth was without form and void so we can understand just how horrible it really was. Ten or eleven thousand years ago was the last major earth axis shift and it quick froze mammoths eating in a field in Siberia when, all of a sudden, the landscape was almost immediately turned into a polar region where everything was dead. The Bible talks about this as being the destruction of the **planet Rahab** and other texts tell us 1/3 of the entire population of the earth was wiped out. A thousand years later another attack temporarily shifted the earth melting the ice caps, forming massive tidal waves and drowning just about everything and everyone left on the earth. When the clamor had ended, the earth shifted back to its 10 thousand year alignment as captured in the mid-Atlantic magma and the Hawaiian hot spot trail and life began again. For a thousand years before the Earth's latest positioning, there was a thousand years of SOMETHING. Scientists call it the Young Dryas. Sometime during this time, Venus was destroyed.

## Venus Was Destroyed

By cross-comparison of information from around the world and from looking at generally disregarded scientific research, this history is like none you have read before. Venus was destroyed by something other than "the Greenhouse Effect" and I'm going to prove it to you. 10 to 15 thousand years ago the people of earth were at war. The results were horrible. The results are horrible. The planet Venus was pushed slightly closer to the sun. Its rotation slowed, its temperature shot to 800 degrees and everything melted. The earth was pulled slightly closer to the sun as well and the water levels rose. Several key civilizations were submerged in the process. We had entered the Holocene Era and Venus simply died.

Venus moved closer to the sun, slowed its rotation to 1/300th the normal rate, all its oxygen was lost, $CO_2$ covered the planet, massive craters along the equator somehow formed, many places on the planet burst open and lava flowed for a short time, all the rivers and oceans were suddenly emptied and whoever was living on the planet we killed almost instantly.

*Preposterous!!!!*

Not only is this not preposterous, it also has a huge amount of evidence and people here in North America to boot with nuclear materials, electricity, and the evidence that nuclear materials had been detonated. This caused the people in North America to suffer greatly and even drove many underground. Let's take a quick look at how all of this played out.

# Before the Pleistocene Ended

Whether you want to believe it or not, kingdom after kingdom became accustom to war 15 thousand years ago. Possibly even some people had even ventured to Venus during this ancient time. Venus would not have been as it is today. It had green fields, huge rivers, massive oceans and air. According to the Biblical account the Planet was called Rahab or "Vain Place". Anyway, the book of "Jasher" indicates that 1/3 of the population of the entire world was lost in a massive World War before the end of the Pleistocene marked by the huge flooding and Noah's escape.

No matter what, we know that something bad happened on the planet Venus very recently [about 10 thousand years ago]. The atmosphere became almost all $CO_2$. There also has been a huge temperature jump to over 800 degrees, the air pressure has jumped to 90 times that of the earth's, and thick clouds of Sulfuric Acid now cover the surface. You can certainly believe that anything that had been alive on the planet was destroyed in the transition. Another noticeable thing is that the rotational speed slowed to almost nothing. The logarithmic chart below shows the rotational ratios to planet size of the other planets. If you a really look closely you will see that Venus is way out of place. It has almost no rotation. In fact, Venus is currently rotating about 1/100th as fast as its sister planet Earth. No other planet in the solar system has had that happen as shown below.

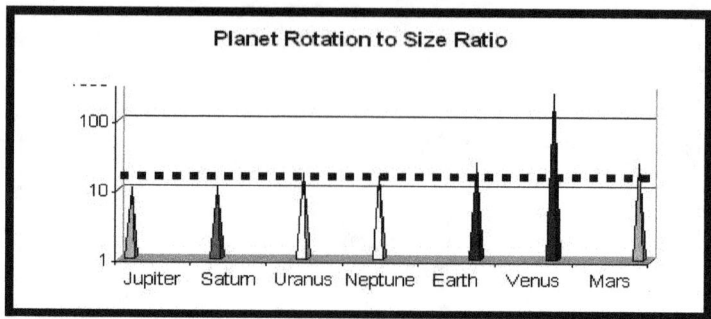

I know some have told you there was too much underarm spray on Venus so it was destroyed by global warming, but there is NO evidence---- let me say that again    NO EVIDENCE. It was all made up!!!!!!!

On earth we had Dinosaur flatulence that would have easily driven the methane gas levels through the roof and major eruptions of magma constituting hundreds of cubic miles of material as has been repeatedly seen would have burned off our environmental shield. The burning of rain forests is nothing compared to the major fires sweeping across entire continents in our distant past. That certainly would not have affected the atmosphere or it already would have. ----Sorry, I just get mad when absurdity takes over common sense. I'll try to contain myself as we go on!

### The End of the Pleistocene

Back on Earth, Noah and others were warned for 120 years that the end was coming. We can assume many prepared in different ways and many ignore the saying. Then something happened on Venus. From physical evidence it appears that its moon was destroyed and peppered the planet Venus and our own planet along the Equator with thousands and thousands and more thousands of massive meteors. It would be too much for Venus and it, essentially, caught fire killing everyone and turning the planet into a massive field of volcanoes which turned the atmosphere to $CO_2$ holding the heat and drying all water.

### Carolina Bays and the End

In North America, something similar happened and quickly halted the world war [I'll explain in a minute]. The huge particles of the Venusian moon blast traveled quickly to the Earth and many hundreds of thousands of massive particles hit along the coastline of North America and around to Australia. Today 500 thousand plus craters up to 14 miles across are still extant. The impact was enough to shift the Earth axis and only those in sealed ships or flying machines survived the massive 500 mile and hour winds, mile high tidal waves, 40 days of rain, earthquakes and volcanic action around the globe. After the Earth settled, huge piles of animals were left to rot and the cities were all destroyed.

Most people probably had an idea about this massive flooding with Noah as a hero, but the next section may not be as well known.

# A Proof of War

Just before this destruction is a time we call the Young Dryas. While Dyra is a flower, the time was no sweet smelling loveliness. Instead it was a wild time and it took a long time; on the order of a thousand years. Imagine a 1000 year war! It was the end of the last Ice Age. Temperatures increased steadily for the next thousand years, but this is where it starts getting interesting. Levels of copper, tin and lead show marked increases and there is this abrupt ending that looks really weird which has a huge drop in temperature as measured in Greenland.

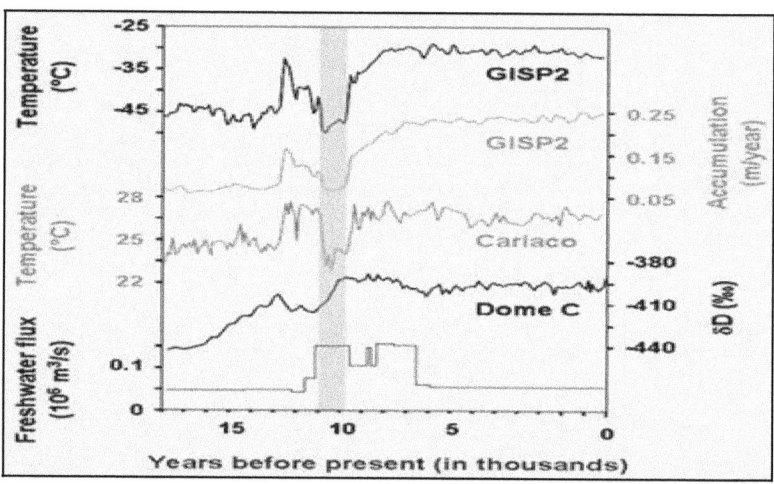

Death was everywhere and then we find something very scary. Uranium concentrations in coral jump by almost 300%. Also we find marked increases in nanodiamonds, magnetic spherules [tiny balls], and carbon spherules at the end of the War with a major increase in charcoal around the middle showing fire unbearable heat and nanodiamonds indicating nuclear explosions confirmed by the uranium concentrations found.----- This occurred during a fairly brief time in our history as shown next as depth can be interpreted as time.

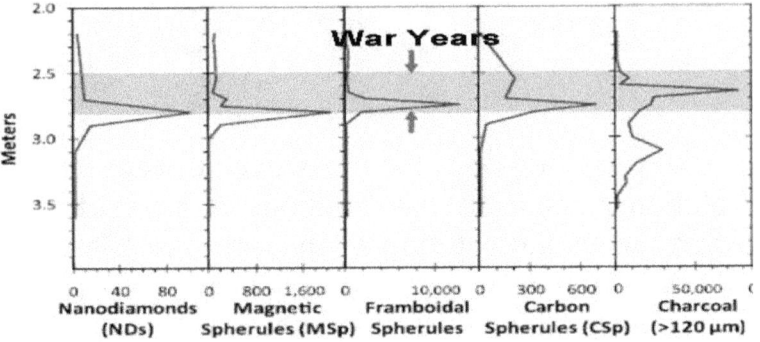

Miles of fused desert sands in Libya and Egypt may also show a massive high temperature "explosion" without a meteor. All the evidence seems to support the existence of an ancient nuclear war having taken place at this time and felt around the world. The huge, unfossilized dinosaurs being found today are radioactive as they must have died during this horrible war. I would tell you about xenon, but I think it would freak you out.

### Xenon-129 Evidence

One way to test for nuclear explosions is by looking for radioactive remains such as Xenon-129. Xenon-129 stuff is a "second order nuclear fission by-product" and guess where too much is found. Mars has nuclear by-product in abundance. Forget I said anything about this, bringing up Venus was bizarre enough. Let me get back to Earth and see what has happened in North America and along the coastline of the United States in particular as this coastline had be along the equator during the war.

# Some Say a Day

We have to worry, not only about overzealous scientist telling us that evolutionary changes take millions of years, but also we must worry about those trying to interpret the Biblical history without cross comparison and reasonableness. You know the ones I'm talking about that think the Earth was made in 6 Yowms [days], when the Biblical history does not say that. Instead the word Yowm mean day like the day of the dinosaurs, or the day of plenty. It simply mean period of time. Here are just a few of the many examples.

*Numbers 28:26* *-the* **day** *of the first fruits [**Actually the first fruits last more than a day so yowm didn't mean an actual day.**]*

*Deuteronomy 20:19* *When thou shalt besiege a city a long* **day**, *[**Laying siege on a city one day, even a long one, would not work so yowm didn't mean day.**]*

*Deuteronomy 22:7* *-thou mayest prolong thy* **day**. *[**Most people live longer than a day so yowm didn't mean day.**]*

*Joshua 3:15* *Jordan overfloweth all his banks all the* **day** *of harvest, [**It takes longer than a day to harvest so yowm didn't mean day.**]*

*2 Samuel 23:20* *-slew a lion in the midst of a pit in* **day** *of snow: [**One day of snow is unlikely so yowm didn't mean day.**]*

*1 Kings 11:42* *And the* **day** *that Solomon reigned in Jerusalem was forty years.[**A day and 40 years can't be the same so yowm didn't mean day.**]*

*1 Kings 14:20* *And the* **day** *which Jeroboam reigned was two and twenty years: [**22 years is not a single day either so yowm didn't mean day.**]*

*2 Kings 3:6* *And king Jehoram went out of Samaria the same* *day*, *and numbered all Israel.* **[An extremely fast census could** **not be accomplished in a "Normal day".]**

*Job 15:23* *he knoweth that the **day** of darkness is ready.* **[Again** **they were not speaking about a single day.]**

*Job 30:16* *the **day** of affliction have taken hold upon me.* **[Job** **was afflicted more than one day.]**

*Psalms 41:1* *the LORD will deliver him in the **day** of trouble.* **[I** **cannot believe that there is only one day of trouble.]**

Just like is says the "day of harvest" or the "day of the king", or the "day of –" simply is a general term for time period and the evidence that the first 6 time periods were very long has been determined by carbon 14, many other dating methods, and common sense. Actually there are a number of issues in the "creationist theory".

# Creationist Theory Problems

Like I said, some try to force fit the absolute time into secular dogma and when there are problems, very interesting devices are used to reestablish the "Absolute Timeline" in what is now known as the "Creationist Theory". Let's look at problems with absolute Biblical timing without regard for other evidence.

The theory essentially is that the world can only be about 6 thousand years old, and the creation of all animals occurred within a 6-day period in accordance with the first chapter of the Biblical book known as Genesis. This very shortened timeline would not allow for any solar system modification, because there simply was no time for it to happen. Later we will see that there is much history on the various planets that could not have happened instantaneously just like there are things on the earth that could not have happened instantaneously. Below are some of the elements that show that the "Creationist time base" is, most likely, way too short.

### *Magnetic Field Changes*

Many Magnetic field changes have been verified by lava samples from the Atlantic Ocean and other sites. As the planet plates separate, lava pours out in abundance in the middle of the ocean, and the metallic portions of the lava align to the current magnetic field of the Earth before it hardens. Core samples from the bottom of the Atlantic show that there have been at least 170 major changes in the magnetic field alignment and none have occurred in the last 2 thousand years. A process was developed which uses these metal alignments for dating purposes. It is called Paleo-magnetic dating and it does not confirm a short time span. Each time that the magnetic field changes, havoc and destruction is rendered on the inhabitants of the Earth. By creationist standards, the 170 shifts would have occurred in 4 thousand years of improbable and unbelievable massive wobbles, destructions, and annihilations and, during all this mess, the Biblical story has almost no mention of the upheavals except for one that occurred 7 days before the worldwide flood. We will discuss in detail the destruction periods that are revealed in the Biblical texts, but they do not account for what is found in the Atlantic.

## *Worldwide Flood*

The worldwide flood could not have created the thick pockets of coal, as implied by the creationists. Even if the flood produced wide and uncontrolled manufacture of coal deposits, the amount and depth of the deposits around the world could not possibly have been generated during that one major Earth trauma. In some areas, the coal deposits are well over a mile thick and are over an extremely wide area. The thought that the trees all congregated in one spot as they floated around in the flood and then collected to form these massive coal deposits is not probable. Trees must have been in an area, then died, grew, died again, grew, etc---for many, many thousands of years.

By the way, there is a huge amount of evidence that there was a worldwide flood, but it most probably happened 9 thousand years ago

## *Physical Similarities*

The physical characteristics between the shorelines of South America and Africa, along with many other indications, strongly suggests that there was once one major continental mass that has been separating slowly over many, many years and it could not have happened over the 2000 years of creation before the flood without showing a major difference in the crustal density in the middle of the Atlantic Ocean. The crustal density is identical in the Atlantic and on the continental masses. Only under the Pacific Ocean is there any major variation in density of the crust, which brings us to another problem.

## Thin Crust

The Earth's crust is thin under the Pacific Ocean. By using seismic mapping techniques it has been determined that the thickness of the crust at the bottom of the Pacific Ocean is much, much less thick in comparison with the thickness of the crust around the rest of the world. This **strongly** supports the premise that the Earth was split apart in ancient times and is slowly healing itself. By examining the amount of crustal matter that is deposited each year, the age during which the Earth was split apart has been estimated to be a little over 200 million years ago **[not the billions of years that many textbooks would have you believe.]**

Many suggest that the current dating methods are flawed and NO reliable timing can be ascertained from radiometry; especially the Carbon14 dating of which we are most familiar. Here are some of the "objections"

## Carbon 14 Anomaly

Without a doubt we have found items that tested to be older than they actually were with this carbon 14 test method, so some say it cannot be trusted. After a volcanic action the amount of carbon 14 was decreased more dramatically than normal. This change shifted the testing results so many say the timing cannot be trusted.  Here is the rub about the volcano and other sampling done where areas were excessively hot. Scientists found items that test samples were dated to be **younger** [not older] than they actually were. This anomaly is caused whenever the amount of Carbon 14 in the test sample was increased unnaturally [usually by enormous heat]. Carbon 14 has a half-life of 5600 years. That type of dating is only good for about 40 thousand years and assumes that no outside force increased the percentage of carbon 14 on the tested sample. Certainly we must look at the environment during the timing cycle or errors can occur, but this type of error is relatively rare and carobn14 is only one of dozens of methods used to identify and verify dates.

## Oxygenated Air Anomaly

It has been suggested that the air was more oxygenated long ago and therefore, the decay process was modified before the worldwide flood, more oxygen should mean more carbon-based items. This would mean that there would be a higher concentration of carbon 14 than we currently see, which would in-turn mean that any carbon-14 dating that crossed the high oxygen boundary would be in error---This error [like the one identified above] would indicate that things that were actually much older would test to be less old because there would be too much carbon-14 remaining. **Items tested to be 6,000 years old may be twice as old, which of course is the wrong direction for the creationist belief.**

### Nuclear Event Anomaly

An additional problem for carbon-14 dating is a nuclear event, and the Earth has seen many, but a nuclear event would increase the available carbon-14. Again there is a creationist dilemma. Increasing carbon-14 means that things are even older than normal carbon-14 testing would suggest.

### Too Many Animals Anomaly

There have been so many animals on the Earth that we still haven't run out of the oil that was produced by the decay of their bodies. If the animals were only here during the 2 thousand years before the flood, the Earth must have somehow been much, much, much larger to allow them to walk and not be piled on top of one another. I know that some believe that oil came from some other means, but right now let's just assume that it came from dinosaurs, because it is the most logical without additional insight.

### Karoo Anomaly

On the southern portion of Africa lies the greatest find of terrestrial vertebrate fossils [mostly swamp dwelling reptiles]. It is estimated that there are **800 Billion;** yes that's billion with a B, animals in a sandstone and shale deposit that is **20,000 feet thick.** It is stretched out for hundreds of miles. This simply could not have been a single clump of creatures pushed into one area as the floodwaters subsided.

## Clumped Animal Anomaly

By the way, I suppose you have noticed that, like the last two examples, oil is only located in small pockets around the earth and huge clumps of dinosaurs have the same anomaly. I just indicated that the floodwaters could not have caused the huge piles, so what caused piles of animals to die together over a very short time period. The answer is huge Comets hitting the earth and periodic, but violent earth axis shifts. Both of these catastrophic events cause an unstable environment. Animals are swept away and twisted together in large piles. You have probably heard about the Mammoth with flowers still in its mouth as it was "quick frozen". This too is caused from one of or both of the aforementioned events. We will look at some of the more dramatic of these events that have happened on the earth over the years. They drastically modified how our planet "evolved" and a shift could happen again very soon.

## Differences Versus Time Anomaly

If one accepts a 6-day creation and a 4000-year time period between the worldwide flood and today, there are too many differences and not enough time. According to Creationist view, the flood occurred 4000 years ago and only olive skinned Adamic people survived. Within a period of about 100 years, they mutated into red skin people, white skin, yellow skin, black skin, brown skin, straight hair, curly hair, flat nose, high cheek bone, and slanted eyed variants around the world. Then, for the next 3,900 years nothing happened at all. Carvings from thousands of years ago and today show people look the same. Those first 100 years must have been something if we are to believe a flood date of 4 thousand years ago and only Noah descendants as the survivors.

# Before the Australopithecus

Please look over the following data before making a critical judgment against this premise. The drawing above shows a normal human and the Australopithecus known as Lucy. Even before these primitive humanoids lived, and before much of our Solar System had settled into a somewhat stable state, the world was inhabited by "normal" humans by all accounts. This is an important element to consider as we go forward so I have provided a very small sampling of the large amount of evidence concerning these ancient people on earth. The following samples are some found in the United States, but evidence has been collected from around the world.

**Oklahoma-1912**-An iron cup was found inside a lump of coal. The estimated age of the coal was cretaceous. [Look at the detail and workmanship from the graphic provided]

 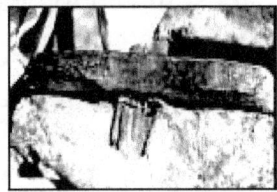 

*Texas 1934-* This graphic is of a hammer with partially coalified wooden handle that was found in Cretaceous Limestone. Here's the strange part. The Iron was processed such that it didn't rust.

**West Virginia 1944-** A "brass" bell with an iron clapper was found encased in a lump of coal. From the picture, you can see that it was very similar to those manufactured today. The estimated age of the coal was <u>Cretaceous</u>.

**Pennsylvania**-Here was found a set of cretaceous human bones embedded in coal, there was found a handle for some type of tool. The handle was coalified just like the sample from Texas.

**Colorado about 1845**-This time it was in the Marshal coal beds some 300 feet below the surface. The "Scientific American" reported that, imbedded in a hollow place in a piece of coal was found a thimble. The coal beds were classed as lignitic. Apparently the thimble was dropped sometime between the Tertiary and the Cretaceous period. **Colorado 1867**-Tempered copper artifacts were found imbedded a silver vein at a depth of 400 feet. Estimated time that marked when the vein was formed was during the Cretaceous period.

**Texas 1976**- Another ancient Hammer was found alongside dinosaur prints. The composition of the hammer was 97% iron. The estimated age of the adjacent footprints was cretaceous. [Just when did the Iron Age start anyway?]

**Nevada 1869**-The remains of a 2-inch long metal screw were found inside a block of feldspar. The calculated age of the stone was Tertiary. The screw itself was completely decomposed, but the rock contained a perfect mold of what had been inside.

**Morrisonville, USA-1891-** A 10-inch long, 8 carat gold **chain** was found encased in coal estimated to be cretaceous.

**California 1877**-A metal Mortar and Pestle was found under some lava beds 300 feet deep. The mortar is about 4 inches in diameter. The estimated age of the objects was determined to be from the Tertiary Period.

**Illinois 1851**-While digging through rock, workers found the remains of 2 copper rings at a depth of 500 feet in the earth.

**Coin in Coal** -A coin-like object embedded in a lump of carboniferous coal was found and was reported in "Strand Magazine" in 1901. The coal would have been formed in the cretaceous period. [No; there wasn't a date on the coin.]

**Pennsylvania 1937** - A woman named Myrna Burdick found a spoon among ash from burnt coal. The ashes had not been disturbed after a large piece of coal was burned, but when they fell apart, the spoon was noticed among them. The coal would have formed around the spoon during the Cretaceous Period.

**1880 *Colorado-*** According to the "American Antiquarian", inside a lump of coal, 300 feet deep, was found a perfectly formed thimble. It was another Cretaceous discovery.

**Iowa 1897-** A large stone [2x2x1feet] with multiple faces of an old man carved on it and a grid pattern on the remaining area was found 130 feet down in a coalmine. The estimated age was Cretaceous.

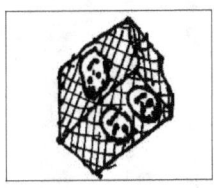 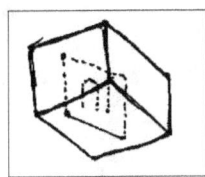

**Philadelphia 1829-** A 30 cubic foot piece of marble was excavated from a depth of 60 feet. Inside the marble was a straight edged rectangular indentation. After a section of the marble was carefully removed it was found that 2 distinct heavily engraved letters similar to an "I" and a "U" eleven inches long and 5.8 inches deep were on a square base. The estimated age was Cretaceous.

**Ohio 1869-** A slate wall was uncovered in a coalmine shaft at a depth of 100 feet. The wall was covered in strange letters. The letters were raised and well defined and the coal around the letters contained the impressions. Each letter was 3/4 inches long and arranged in lines of about 25 letters. The estimated age was Cretaceous.

## Strange Geodes

The picture to the left below is some type of power conversion device found **inside** a geode, in California. Below the geode is a drawing of x-rays of the geode showing the elemental parts. These include a spring, core, plate, and electrical insulator; the same parts you would expect in a battery. Maybe this is a new way to package batteries, but it takes a long time to complete the package. Both of the objects are extremely ancient and certainly before we originally thought that everyone used electricity. The central metal core surrounded by the white material looks like a battery, although some even attribute the structure to that of a spark plug. Whatever it was, it was electrical. On the right is a drawing of the parts and a size comparison to a standard D-cell battery.

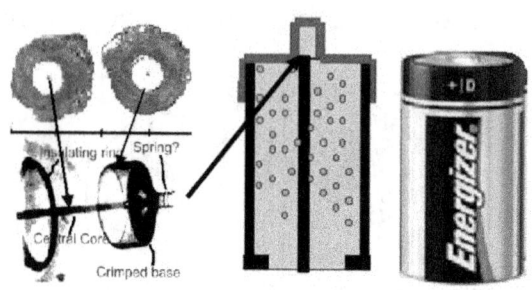

## Another Geode

If that wasn't strange enough, still another geode was found with more goodies left behind by the ancient humans. Geologist Mike Walters found this geode. Inside was a worked metallic bar or ring. This could only have been man-made.

**California Find**-Inside a Piece of Quartz crystal, a small metal piece was found it was obvious that it had been worked and was in the shape of a bucket handle.

*Amargosa Desert 1961-* On the edge of the Desert was found a geode. Inside was found an iridescent stone with a 2mm x 17mm long **metal rod**. [You know Geodes must be old.]

## DNA

I'm bring these up to let you understand that a civilized group of people live in North America well before the end of the Pleistocene Age and well before that land bridge theory took place. It should be noted that Haplogroup DNA ancestry testing has shown an anomaly in the Americas. The A, B,C, D and X mutations are common in the Americas and are isolated in the Euro-Asian continent. That seems fine, except there are no trails of these mutation groups across Asia to where that land bridge thing would have been. Instead they disappear and reappear in North America by magic [or flying machines]. Somehow they popped up here before the end of the Pleistocene age as the mutation dates are about 12 thousand years ago.

So now we have a base to start. Mars yanked out the Pacific, Venus peppered the North East coastline with meteors as it was destroyed; massive nuclear war was evident around the earth and the earth shifted 10 thousand years ago sending the Mammoths to their immediate death and flooding the world. If we back up a little we can look at the planets.

# The Solar System Begins to Stabilize

I've told you about anomalous apes being eaten by dinosaurs and ancient, civilized humans living and working on the earth to get you thinking that you might not have been told the whole truth. Just like those things, the steady development of our Solar System doesn't match the evidence.

### Forget Bode's Law

People have been trying to force fit the planet positions into a neat package called Bode's Law for years without success. It was invented by a man named Johann "Titus" in 1766 so we got the bright idea to call it Bode's Law, but it is not even close to being a law. Some of the planets seem to be close, but when it comes down to it, the positions don't match. Our Solar system is still forming. I know it's hard to believe, but even as recently as 11 thousand years ago there appears to have been a major repositioning of the Bode's Law controlled masses. The best thing we can do is to simply ignore it. **Cross it out of your science books**. Here are some of the time-stamped diagrams of our solar system through the ages. I'm mostly going to concentrate on Mars, Venus and Earth, because they are the closest together.

### Born 4.55 Million Years Ago

As I explained, the solar system generally was established 4.55 million years ago and with it, the planet Earth. The ancient texts talk about the planets being fluid; especially the earth. New evidence suggests that some of the planets were in a state of flux as recently as 11 thousand years ago, but that will have to wait as we take the first snapshot of time some 2 million years ago.

You were probably told everything was stable by this time. The spinning gases that made up the various planets had compressed into solid masses with almost circular orbits around our solar system star. Unfortunately for those who were supposedly being taught, the orbits were not stable.

## The Obstinate Planetoids

Many of the planetoids did assume their place very soon after the solar system came into being, but some were obstinate. Of the obstinate ones, earth, the moon, Mars, and Venus were the ones most obviously changing.

By the way, thousands of planetoids sort of jumbled around and essentially orbited the sun in what is called the Oort cloud just like they do today. There is evidence that this particle filled cloud might have been much closer to the sun in the olden days, but most of the particles had little to do with the terrestrial planets except for one simple fact. During the ancient times more comets were generated which meant that more significant meteor strikes occurred on the planets which wreaked havoc somewhat more often than today. Below is a drawing showing the general relationship of the various planets. I purposefully drew earth farther from the sun than Mars, but we will find that their orbits were radical and they took turns being closer.

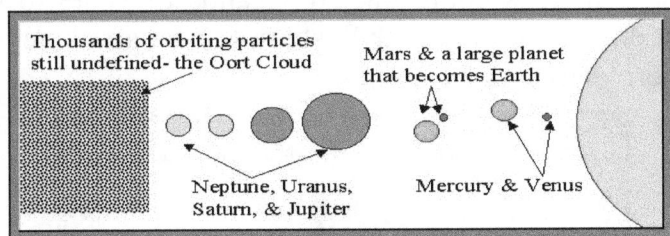

## Solar System Changed

Current evidence strongly suggests that the Solar System changed about 250 thousand years ago. I know you were told this was much earlier, but there is evidence to suggest the much more recent date. A super planet close to Mars may have finally exploded from the gravitational pressures and the asteroid belt was established. The <u>largest part probably became the Earth</u>. I know that concept seems fanciful but please continue reading and its probability will become much more understood. Be patient.

Beyond not believing this event occurred ourselves, we have a very difficult time believing that this event would be understood by ancient man, but many historical references are almost identical in the details of this event and new computer simulations of this event help confirm some anomalous characteristics of the solar system of today. You may not believe these facts right now, but some of the proof is described as soon as the next chapter.

### Oort Cloud and the Kuiper Belt

The Oort cloud also split around this 250 thousand year old time frame. At least 31 on the planetoids established a somewhat more stable orbit around the sun. The more stable area is now called the Kuiper belt. One of the largest planetoids in the group is named Pluto. Exactly how it got there is unknown. According to interpretations of the Sumerian texts by Zecharia Sitchin, Pluto was a moon that was pulled away from one of the Jovian planets as another extremely large planet entered the "known" Solar System from its eccentric orbit in the Oort cloud. I'm not getting into additional planet theories in this work, but certainly it is possible to have large masses with eccentric orbits in the Oort cloud. These masses could take thousands of years to orbit the sun. Upon entry into the Solar System, the rouge planet could be very dangerous to smaller bodies

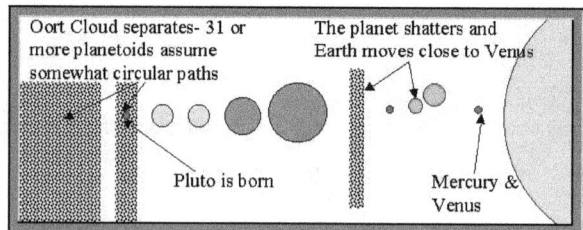

### The Solar System Changed Again

About 11 thousand years ago the evidence suggests that another interesting change occurred. The moon of Venus exploded which sent thousands of particles raining down on the inhabitants of the Earth and set up Venus for a terrible disaster from which it never recovered. The planet earth didn't do so well during this time either. Aren't you amazed that I added the Venusian moon. That discussion is much later in the book.

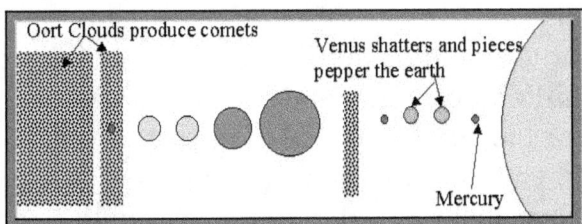

Until you read the evidence, don't simply ignore the very recent change in our solar system, because there is a lot of evidence to support the theory and this change really affected modern man. You'll see, but let's first see how the earth was "modified" 250 thousand years ago and the ancient writing that support the event along with the scientific evidence that has been collected.

# Creation Stories

We're back at the time of the first depiction of the solar system around 4.5 million years ago and our investigations must turn to the creation of the Earth. For the first step, we need the help of ancient texts and stories from around the world. You'll be amazed at how many of the creation "myths" are strangely similar. Here is something we need to keep in mind. If everyone remembers details the same, then there is a much higher probability that the information is correct. At least it is as correct as the people writing the history could understand.

**Similar Creation Stories**-I've put together 17 substantially similar creation stories from around the world. Everyone told the same story. From one culture to another, the details are so close that it's almost impossible to ignore that they are, basically, the same story. Their similarity is not dependent on closeness of culture or placement of the people. In fact, a common theme in these historical references is the concept of the Earth splitting in two. Half became what we call the Earth and a firmament was made from the other half of the planet. What on Earth would have given the ancients such a strange concept? From these excerpts one can begin to get an understanding of what might actually have happened or what a large segment of the people believed happened, millions of years before the time of the stories being told. Greek, Sumerian, Aztec, Egyptian, Chaldean, Hindu, Ute, Maori, Celtic, Hopi, Chinese, and Polynesian people have no basic history together, but they all have the same insight and somehow had knowledge of what must have happened when the Earth was formed. I've tried to keep the overviews short so that the basic concepts and similarities can be quickly seen. Afterwards I have presented a couple of interpretations to consider given these similar stories. Don't throw data away just because it is strange. When trying to examine the past. USE the stories of the past!

### *Greek Creation Story from "Argonautica"*

*The first being was Gaea. She bore Uranus and then married him. They were the first to rule Olympus.The birth of Gia & Uranus's children resulted in a war by the gods that lasted for generations. [**Like many of the ancient histories, the heavens were in a state of flux or "at war with one another"**]Gaea and Uranus fashioned the Cosmic Egg, which contained the earth, sky and sea. It exploded and the earth, sea and heaven were separated. [**Just like almost all creation texts, the Earth split apart in an ancient age.**]Next, the Mountains rose from the sea. [**If this is talking about 200 million years ago it was when a Mars flyby caused many of earth's mountains to form.**]Then Gaea and Unanus populated the earth with animals. Their children, the Titans [Rhea and Cronus] took over.During his reign, Cronus created mankind. This was the Golden Age. The most peaceful time of mankind.*

### *Chinese Version-from Chinese legends*

*The first living thing was P'an Ku. [**God**]Almost identical to the Greek version, he evolved inside a gigantic cosmic egg, which contained all elements of the universe. The sky and Earth were one. [**Sky and Earth was a super planet**] He separated the Sky from the Earth. Gradually he separated wet and dry, Then the light and dark. While he grew he also created the first humans.*

### *Hindu Version*

*Brahama was the progenitor of the whole world [**The "whole world" was the "heaven/Earth" super planet**]. After a while, the divine one, by his thought, divided it into two halves. [**Earth split**]From the two he formed Earth and the firmament [**asteroids**]. The eternal abode of the waters was made,Then he created all beings.*

### *Central American Aztec Version*

Originally the Earth and Sky were one named Coatlicue **[Coatlicue was the "Earth/sky" super planet]**The creator ripped her into two pieces forming the sky and the Earth. After the creation, the gods saw that the Earth was formless. Only the ocean was everywhere. Quetzalcoatl dived into the ocean and killed a large monster that was under the ocean and formed the land from its remains.

## Egyptian Version

**[Ffrom the Egyptian Book of the Dead]**-The God Re created everything including Ra, the sun god. Then Ra got himself pregnant 1/4 of his offspring became the Earth and three fourths became the sky. **[Ra's offspring combination was the "Sky/Earth" super planet]**Light streamed forth, banishing darkness on the third day. The great "shining ones" were born on the 5th day. Last he created man and the other creatures on a potter's wheel and breathed life into them.Men were so wicked that Ra had enough and decided to kill all mankind. **[Flood]**

### Digueno Tribe Version [Africa]

In the beginning was Tu-chai-pai. He made the world. The male part was the heavens and the female part was the Earth. **[The male and female part was the "heaven/Earth" super planet]**Tu-chai-pai said "We-hicht" three times, which caused the heavens to rise above the Earth. **[Earth split]**First he made the ocean and the land, He then made man out of mud. Man was easy to make, but woman took a long time. Then he yelled, "People, you can never die!"

### Nkongolo Tribe version [Africa]

In the beginning was the Creator. He made man immortal. **[After many millions of years we find less and less evidence of the ancient humans—Some believe they had found a way to become a "spirit"]** The Earth and heaven were created together. **[This was the "heaven/Earth" super planet]** The Earth and heavens were separated, **[Earth split]** When this occurred, man *lost his immortality [**A new type of man was created just like many of the Creation stories.**]*

## Thialand/ Loas Version

*In the beginning were "the three Great Men" [**God trinity**] The heavens and the Earth were joined together by a rattan bridge [**The joined heaven and Earth was the "heaven/Earth" super planet**] The great men placed the Thens [**angels**] to rule the heavens. The people refused to worship the Thens. The Thens overwhelmed the world with a flood*

## Polynesian Version

*The supreme God Io, created the world [**This world was the "Earth/sky" super planet**]. In the beginning there was only waters and darkness. By his word and thought, Io separated the world into the Earth and sky. [**Earth split**] He said "Let the waters be separated, Let the heavens be formed and let the Earth be."*

## Maori Version [New Zealand]

*Tane the Creator God made the Earth and sky as one [**This was the "sky/Earth" super planet**]. The world was full of darkness.*

*Tane forced the sky away from the Earth. [**Earth split**] Light [life] came into existence. The Earth cried and cried and flooded the world.*

## Celtic Version-

*God made the Heaven and Earth as one [**This was the "heaven/Earth" super planet**]. Titans were on the World. One of the Titans cut them apart and split the heaven part into many pieces [**Earth split and asteroids were made**] This allowed light to shine on the Earth. A giant flood occurred from the split blood and all humans were killed except two.*

## Northern Europe Version- [from "Poetic Edda"]

*First there was a huge giant named Ymir. [**Ymir was the "Earth/heaven" super planet**]. The trinity of Gods killed Ymir* *and he was split in two. His body became the Earth and the head became the sky*

### The Hopi Indians- [from the "Book of the Hopi"]

*The creating power made the world before this one **[This original world was the "Earth/sky" super planet].** He was displeased with the people. He stomped on the world. It split in two **[Earth split].** Water was forced from the cracked world. Mud was brought up and became dry land. God wept and his tears became the ocean. Then he created man **again.***

### N. American Snohomish Indian [Tradition]

*The creator deity named Dohkwibuhch, created the world. Dissatisfied with the world, a wise man discovered that everyone bumped their heads against the sky, as it was very low. **[The earth and sky were originally together**.] Some people climbed up high trees and went into the sky world. **[There was communication between planets and earth in ancient times.]**The wise man said that if all the people and all the animals and all the birds pushed at once, the sky could be lifted away from the Earth and it was so. **[The earth and sky were separated.]**Some people who did not know of the sky-lifting, were hunting elk and followed the game up into the sky, and were stranded there forever. **[Some ancient humans traveled into space.]***

### N. American Ute Indian Version

***[See how very similar the Ute Indian and Sumerian versions are.]**The sun god, Ta-wat, and the god, Tavi, were in the beginning. **[Tavi was the Earth/heaven super planet]**Tavi came so close to Ta-wat. **[War in Heaven]**Ta-wat shot an arrow at Tavi's face. **[God captured the rebel angels and the created creatures]**Tavi was slivered into 1,000 pieces **[Earth split and Asteroids were made].** All that remained was Tavi's head **[Tavi's head was the Earth].** The eyes of Tavi finally burst from the heat. Tears gushed out and flooded the world.*

### Middle Eastern Chaldean Version- [from "Berossus"]

Belus was before [**the creator has always been here**] Belus cut Omoroca in half. [**Omoroca is the "heaven and Earth" super planet**]Half became the Earth and the other half became the heavens. At the same time Belus destroyed the animals and the abyss. [**The first creation was destroyed**] He then cut off his own head and created man from his blood to give them divine knowledge. [**He made man in his own image**] Then he divided the darkness. Later Cronos appeared to Xisuthrus [**Noah**] and let him know of a great flood coming, to bury historical records and build a ship to hold family members and all species of animals. [**Flood**]

## The Zoroaster version – from "Zadspram"

In the beginning Ohrmazh was the light [God]The heavens were made, the sea was made. The sea was scattered which animated the Earth [**Life was formed**].The Earth was bound together-[**The Earth was the "Earth/firmament" super planet**] A great rain in the beginning of creation tore the Earth by noise and wind. One portion, as much as ½ the whole Earth remained in the middle and the other portion formed the ocean [firmament] around it. [**The Earth split and asteroids were made.**]

### Sumerian/ Babylonian Version

[In the Sumerian version Omorka was called Taimat, but most of it was identical.]The gods Marduk [an outside planet] and Omorka were created. [**Omarka was the Earth/firmament super planet-**] Omorka ruled over the "men with wings" and beasts created by chaos [**Men with wings could represent flight into space.**] Marduk came too close to Omorka and disturbed Omorka. [**There was war in space**] Marduk unleashed an evil wind at her face and captured the creatures. [**God [Marduk] won.**] He sent an arrow into her belly. It split Omorka in two. [**The Earth split**] The half of her was set as a screen [**the firmament {asteroid belt} was formed**]. Omorka's skull was cut loose. [**A new Earth was formed which went to a new orbit**] The Lord raised the flooding storm. From the blood and clay, humans, animals, stars, sun, moon and everything else was created.

Hopefully, you interpreted a level of similarity between the vary histories from around the world. It seems that everyone in the ancient world believed that the earth was once much larger and separated. It is fairly doubtful that if this "separation" had happened 4 billion years ago that it would not have been so catastrophic as to have been the main them of the ancient religions. Relying on ancient texts to form an opinion is dangerous because interpretation, translation, and belief in supernatural could taint the historical nature of the texts. We need more. The first "more" comes from the moon.

**Our Moon is Too Big-**All other moons that were either made from captured particles or debris split away from the host planet are small in comparison to the host planet. The Earth's moon, however, is over 10 times as massive as "almost all" other planets moons when compared with the mass of their host planet. How can the anomaly be rectified? Zecheriah Sitchin, a well-respected Sumerian historian and investigator, has interpreted the Sumerian texts over many years and his detailed analysis indicated that the Sumerians believed that the moon was originally the moon of a much larger Planet named Taimat. He further interpreted texts which indicated that the moon left Taimat and went with the Earth.

If Taimat was the super earth planet before the last Mars/earth close encounter, then the moon probably came from the Pacific Ocean.

**The Old Bode's Law Again-**I know I said to throw Bode's Law away, but should there be a planet where the asteroid belt is as they bogus law suggests? Mr. Sitchin's research indicated that, according to Bode's Law and other data, there is strong evidence to indicate that when the solar system was first "created", a planet was originally in the place where the asteroids are and the Earth was not in its current orbit. Something must have happened to change the original solar system. A planet was destroyed that was where the asteroids are now. As I said before, Bode's Law is no law, it simply shows reasonable comparison between planet placement with respect to the solar system and it may be a reasonable tool when that is understood. Speaking of understood, do you understand the first chapter of Genesis. While I do not profess to know all, the thing I want to do here is show you that the Biblical history AGREES with the histories previously brought up.

# Biblical Version

The Bible seems to say the same thing that the earlier stories did. Before you get mad, let's go right to the book of Genesis and see if it says anything about a super planet coming apart during ancient times.

### The Super Planet not Two Things

*Genesis 1:1-First God created the "heaven and Earth" thing.* **Heaven/Earth Mistranslation-** We know that it was neither the earth of the heavens as they were created later. If you are thinking this "heaven" was where angels live, you are wrong again as that place was created later as well. This is identical to the Earth/heaven super planet identified previously.

### Explosion and Separation not Designed as a Void

*Genesis 1:2a-Then the Earth* <u>became</u> *without form and void.* *[The became is interesting in that the earth must have originally had form. This sounds like a massive war that destroyed everything, just like that mentioned by some of the previous histories.] Genesis 1:2b-Then* <u>Darkness [also translated destruction]</u> *was on the* <u>face</u> *of the* <u>deep</u>*. [A more easily understood translation might be "Destruction was seen in the deep."—I'll go with deep space. The Earth & Heaven super planet was split apart.]* **Darkness Misunderstanding-**The word choshek not only means darkness, it also is commonly translated as <u>destruction</u>. **Deep Misunderstanding-**The word "Deep" cannot be translated as ocean or sea- completely different word. **Face Misunderstanding-**The word "paniym" not only means face, but quite often means look or see.

### God Creates Life not Light without a Sun

*Genesis 1:3-Then God said let there be light.* **Light Misunderstanding-**This "light" certainly wasn't the seeing kind. After all, the sun had not been created at this time. The more reasonable translation from the Hebrew is "life". Life was created in this verse.

## Firmament not Water Canopy

Here is one that gets some people confused so we need to look at the verse, identify its similarity with the other creation texts presented, look at confusing elements and describe evidence that shows that this particular verse is not talking about the water covering in the atmosphere of earth as some have suggested. We will find that "firmament" describes the same asteroid belt separation that the others seem to describe. We will also recognize that "waters" which really is interpreted as "life giving places" doesn't only mean "life giving oceans", but also "life giving planets".

*Genesis 1:6-And God placed an angel-guarded firmament [barrier] between groups of waters [**The more easily understood translation might be the asteroid belt was placed between the two groups of planets, Jovian and Terrestrial.**]* **Waters Misunderstanding-**Waters is certainly not water you drink. The more reasonable interpretation of the Hebrew is "life giving place". This sounds like a planet to me and its confirmation can be found in Sumerian accounts. That's how the firmament, probably the asteroid belt, could separate waters below and above. The asteroid belt was made up of pieces split away from the new planet Earth. It separated and still separates two groups of planets.

**Water Canopy Misunderstanding-**Let's begin by assuming that there was a barrier that separated huge amounts of water outside the atmosphere and huge amounts inside. One would have a hard time believing that the sky could be considered as a barrier, but an even more difficult thing to imagine is that this water outside our atmosphere was not dissipated into the solar system for thousands of years before it became the major part of the water experienced during the worldwide flood.

**Oxygen Level Interpretation-**To prove this theory one may hear about how oxygen levels found in trapped air inside prehistoric amber are found to be at a much higher concentration than today's air and the air pressure was higher so water was in a canopy in the sky. The high pressure and high concentration of oxygen do show a major difference in the ancient sky. Higher air pressure could also be used to eliminate the ridiculous notion that flying reptiles could not fly as we are continuously told. These little clumps of sap, unfortunately, don't tell the whole story.

**Carbon 14 Debunk Dilemma-** Carbon 14 is made whenever cosmic bombardment of neutrons hit Nitrogen14 to produce the unstable carbon and hydrogen. This reaction goes on all the time and at a fairly constant rate, so the amount of $C_{14}$ remains a good tool for tracking dates. If you put a blanket of water around the earth, the cosmic rays could not penetrate and less of the timing agent would be generated. This would make everything seem older than it is tested to be. It could also allow animals to live longer because of the reduced cosmic bombardments, so the canopy theory seems to explain flying dinosaurs, long lives, and inappropriate dating methods. The problem is that most of the other dating techniques are not affected by less cosmic energy and they provide us with similar dates during cross comparison. Therefore instead of finding a vehicle to prove a canopy and timing anomalies, the consistency of Carbon 14 dating seems to indicate that if there was a canopy, it was gone over 100 thousand years ago, the general limit to accurate assessment with $C_{14}$. Of course that by itself could be easily biased and we should examine further.

**High Temperature and High Pressure Dilemma-**Studies have tried to test the canopy limits. First the easy measurements would be noticing that Mt. Ararat is 5 km high and Mt. Everest is almost double that at 9 km high. If we know that the canopy was a vapor instead of a liquid so that the sun, moon and stars could still be seen, the density would by $1/18$ that of liquid water and a blanket of water vapor would be well over 150 km thick. In order to make all that water an invisible vapor, the temperature on earth would have been much, much higher than today and the pressure would be equivalent to being over 9 km underwater or about 13000 PSI instead of our comfortable 14.5 PSI. There would not be many animals that could have survived such pressures; especially not the huge dinosaurs. I know that many feel that ½ the water came from underground lakes, so let's just say the pressure was 7000 pounds per square inch. People, animals, and plants would still be in serious trouble. They would not be able to breathe even with the oxygen content of the air at a higher level. This is not even getting into what the temperature was like to boil all of that water in the atmosphere into a gas. The earth would have been too hot for any creature to live.

## Confirmation of the First Section

After verse 8, we begin to get confirmation of what the first section was trying to say.

*Genesis 1:9-Then God made the dry land to appear in the midst of the water on Earth.* **[In this verse the writer reverts back to the word water instead of waters so it is addressing the ocean. It confirms a difference between the two words.]**

*Genesis 1:14-After that, God made the Sun and moon to shine.* **[Now we get to the light from the sun. This confirms the fact that the first indication of light was not the visible kind but more probably was LIFE. ]**

*Genesis 1:27-Finally he re-made all living things and man.* **[Man was "RE-MADE" to "replenish" the earth. This confirms the fact that the earth had been destroyed. By the way, it also confirms the fact that humans and animals were on the earth before he remade them during this time period.]**

# Physical Evidence of Planetary Creation

This section is not on the beginnings of the Solar System. As such the planets had generally formed and were spinning around the sun. Our text books give us a glimpse into the beginnings and there is no major reason to embellish the extremely distant past. I needed to move the 200-300 million year time frame a little as I previously brought up, and I needed you to understand, in the very distant past, humans lived on the earth. The reasons for involving these ancient humans will become more apparent as we look at some of the other planets. Things seemed to be going fairly well for the earth until a catastrophe beyond belief changed everything. There had been many destruction periods and many life changing events, but this one was worse than any that the humans of that time could have possibly remembered. The destruction occurred about 400 thousand years ago according to a number of the dating methods I previously mentioned.

## Earth 400 thousand Years Ago

Just because someone writes a story like those previously addressed doesn't mean that is what happened. However, it should be noted that writing fantasy for the sake of writing doesn't make sense during this time because writing was simply too difficult. That being said there is a tremendous amount of evidence to show how the earth exploded and the asteroids were put into place, and life was remade after a great catastrophe. All of these things apparently happened around the same time. That time was 400 thousand years ago.

# The Wrong Model

To expand on this theme and to show perspective in the solar system we need to first look at the solar system. The drawing below shows the mean revolution distances of the planets in a linear scale. Some have indicated that they have a formula that positions the planets this way, but the formulas always fail. There is a marked difference between the terrestrial planet distances and the Jovian planet distances.

The creative mind came up with Bode's law. In it an arbitrary distance of .15AUs was used as the base and the distance was arbitrarily doubled and then a 0.4 AU distance was added to each number. The model was not great, but by picking the right arbitrary number, this is what was shown.

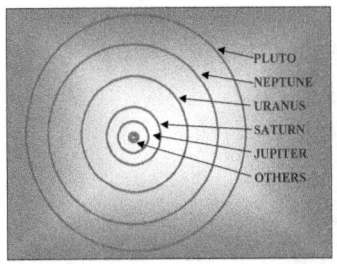
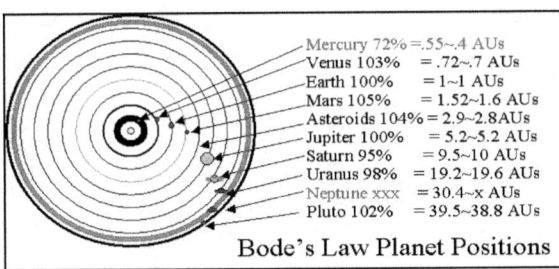

Bode's Law Planet Positions

Mercury was all wrong and Neptune was not supposed to be in the solar system at all, but it looked sort of right. The universe does not run with seeded and arbitrary numbers, however. Almost everything uses something  called a natural log, so let's see what the solar system looks like with a natural log base.

## Natural Log Solar System Model

The  natural log solar system is shown next. What we see is that Neptune, again is all wrong, but more to the point, Mars and Earth have both been shoved closer to the sun than would be predicted.

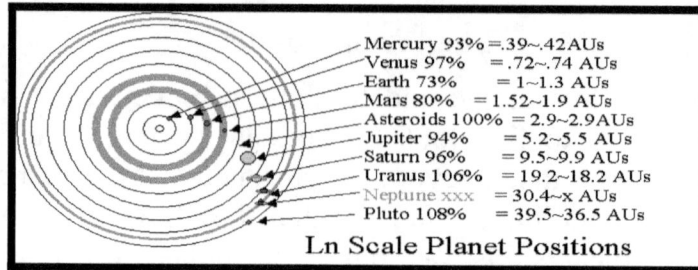

| Mercury 93% | =.39~.42 AUs |
| Venus 97% | =.72~.74 AUs |
| Earth 73% | = 1~1.3 AUs |
| Mars 80% | = 1.52~1.9 AUs |
| Asteroids 100% | = 2.9~2.9 AUs |
| Jupiter 94% | = 5.2~5.5 AUs |
| Saturn 96% | = 9.5~9.9 AUs |
| Uranus 106% | = 19.2~18.2 AUs |
| Neptune xxx | = 30.4~x AUs |
| Pluto 108% | = 39.5~36.5 AUs |

**Ln Scale Planet Positions**

This "new" view of the planets is more probable than the linear model or the stupid Bode model and it makes sense. Everything looks different in this scale, but it looks much more uniform as the planets are slowly establishing reasonable placements. One thing to note is that you don't have to disregard the terrestrial planets for being too close together in this scale. Many of the elements line up in  orderly, evenly distributed rings, but the following should be easily recognized.

- *Neptune's and Pluto's orbits are almost on top of one another in this scale*
- *There is too much space between Jupiter and Mars*
- *The orbit of earth is too close to the sun.*
- *The Asteroid belt lines up with a location of a forming planet.*
- *Those things can be more easily explained than the gyration necessary to explain the "linear model".*

**Neptune/Pluto Problem**-One of the two is not a planet. I'll go with Pluto. Also it should be easily asserted that something has pushed Neptune in closer to the sun at some time in the development of the solar system.

**Space Between Mars and Jupiter**- There was something there at one time.

**The Terrestrial Planets are bunched up slightly**- Erratic motions of the early Martian and Earth orbits caused Mars and Earth to come close together, the ensuing explosion and loss of planetary material pushed the two planets closer toward the sun. and Venus and Earth had an encounter at a much later date. The graphic following shows the exchanges predicted by mathematical simulations.

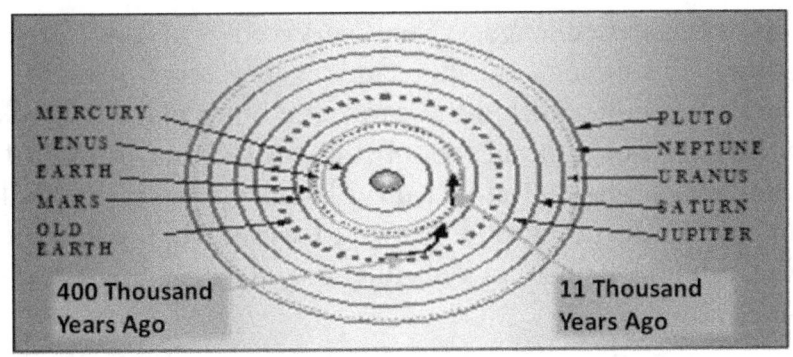

| 400 Thousand Years Ago | | 11 Thousand Years Ago |

MERCURY
VENUS
EARTH
MARS
OLD EARTH

PLUTO
NEPTUNE
URANUS
SATURN
JUPITER

**400 Thousand Years Ago**

**11 Thousand Years Ago**

Somehow, humans and a few animals survived the catastrophe of 400 thousand years ago and the only remaining continent, typically called "Pangea" began to split up. The Continent of Prestonia was thrown into space. Again humans seemed to have everything under control and WHAM! Another killing event. The time was 120 thousand years ago.

## Earth 120 Thousand Years Ago

While the planetary obits may not have changed during this time, there can be no mistake that, at least on earth, something terrible was going on. There is a great amount of evidence that suggests that the earth was spinning at a much higher rate than it is today. While one of the initiators of this event was a set of meteoric collisions, the total reason for this change may never be known. The event not only killed many inhabitants immediately, it changed the equilibrium that had caused the rotation to be somewhat constant. The slowing down of the earth's spin changed everything. Man survived again. In fact, only the extremely large animals could not live well in this new world. Humans thought everything was going along fine again when WHAM! Several more catastrophes happened. The one of interest to the solar system happened about 11 thousand years ago.

## Earth 11 Thousand Years Ago

There is a great amount of evidence about this catastrophe. Biblical and other historical accounts indicate that 1/3 of all life was gone. Worse than what happened on the earth was something that changed Venus forever. That will have to wait until we get through the earth changes. Man survived, but not for long. Something happened 6 thousand years ago. While this catastrophe greatly affected the earth, the other planets were also affected. I'm not talking about the worldwide flood of 10 thousand years ago that ended the Pleistocene Age, I'm talking about a great war 6 thousand years ago. Strangely we find most 2 position mutations in humans either occurred at this time or 6 thousand years ago.

## Earth 10 Thousand Years Ago

There is a great amount of evidence about this catastrophe. Biblical accounts and over 700 other ancient texts recorded this important event in earth's history. A worldwide flood destroyed almost everything.

## Earth 6 Thousand Years Ago

We find something strange here as I just mentioned, massive mutation of the Human DNA occurred at this time and modern haplography techniques have been used to date the mutations. After the earth had dried out a little, the humans of that day almost immediately began to prepare for more war with others in the heavens. Maybe it was to insure that another flood would not destroy so much, or maybe it was simply some crazy notion of Manifest Destiny. Who knows. Around the world we find the evidence of mass destruction. From ancient Indian texts we find that the war sped to the Lunar colonies, so we might believe that other planetary colonies were equally affected. By the time the war had ended, there was destruction everywhere and the people on earth had begun living underground. Each year we find more and more of the underground cities and evidence of this very unstable time for earth and the surrounding planets.

## Timing the War

The Bharata War, was one of the names of this horrible world war and it was described and critically timed by the Maya, the Brazilian Mongulala, Egyptians, Hindu and astrophysics as described below.

*Before 2500 BC- the ending war according to Palermo Stone from the 5th dynasty Egypt*

*3000 BC-* *according to Dr. B.N. Narahari Achar planetary software and astronomical references in Mahabharata*

*3066 BC-* *according to Dr. D. Abhyankar subtracting 38 war years of Mahabharata*

*3067 BC-* *from Planetary software and description in the Raghavan*

*3090 BC-* *Median date for Egyptian Zep Tepi [New Beginning]*

*3100 BC-* *according to Dr. N.S. Rajaram (astronomical statement and interpolated passages of Mahabharata)*

*3100 BC* *according to the reunification of the upper and lower Egypt after the war years*

*3104 BC-* *according to the start of the Age of Kali [Hindu new beginning]*

*3114 BC–* *according to the start of the Mayan Calendar-new beginning*

*3127 BC-* *according to the Aihole Inscription of 7th century AD Egypt)*

*3143 BC -* *according to Shri P.V. Holey Astronomical measurements of Mahabharata)*

*3400 BC-* *Mongulala historical reference- end of the Blood Age [Brazil]*

Just like the World War during the Pleistocene Age, this one destroyed civilizations around the world. It would thrust people back into the Stone Age and so much nuclear fallout or biological weaponry was used that massive DNA mutation changed the course of history and the development of mankind.

Since then we have been playing a waiting game. When will the next major catastrophe occur? Will it be manmade or natural? I'm not talking about a tsunami, hurricane, of volcano, I'm talking about destruction. To give us perspective and insight, let's look at the details known about the 400 thousand year old destruction period. As I briefly explained before, the evidence indicates that Mars did it.

# Mars Molded the Earth

The name of this section seems strange at first, but the evidence confirmed by the previous ancient texts indicates that some major planet was instrumental in forming the earth. We will see that, in all likelihood, Mars molded the Earth. The drawing below show Mars coming close to the earth and exploding. Note that there is a continent on the earth where the Pacific Ocean is today [I call it Prestonia]. I know you have only been told about Pangea. It supposedly included ALL landmass on the entire earth, but that is simply absurd. The earth's spin would not allow only one side to contain dry land.

Notice in the graphic that Mars had water and an atmosphere [as indicated by the blue and green] before the event. Notice also that the Moon came from the explosion. Finally, notice that the Earth and Mars were pushed into different orbits from the incident. The story needs explanation and confirmation, so here goes. For the explanation we look at plate tectonics.

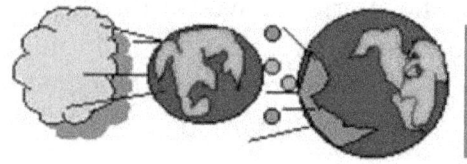  

### Plate Tectonic Anomaly

I mentioned this before, but for a minute, let's go back to school. From the time you were in grammar school all the way through college you were continuously told that "plate tectonics" formed the mountains of the Himalayas, reasonably called the Himalayan Ridge. This same thing caused the long range of mountains called the American Ridge that cover the western side of South and North America. We were told so many times, that we didn't question its absurdity. The plates simply went past each other and all the dirt and gravel that was on top built the mountains. If the Himalayan ridge is the intersection of two "Plates" and one plate was rammed against another huge mass about 400 thousand years ago such that **"quadrillions of tons of Earth"** were pushed up an average **"2 miles into the sky"**, it took more than a few little volcanoes to push the slab with the force required. It would have taken a cataclysmic event we can't even imagine. Then, supposedly, the same thing happened again for the American mountain ridge.

Estimates have been made assuming the event initiator was volcanic. If both mountain ranges used plate tectonics and the massive pieces of earth's plates were driven by volcanoes, **the two "volcanoes" required to do this would each have been about the size of the United States**. I don't know how much energy would be expelled from the mother of all volcanoes, but what I think the researcher was trying to say is that it was impossible. There just isn't any evidence of these huge explosions and I think if there were energy producers available, to move the plates, the earth would have more serious problems than the mountain ranges being formed. The whole concept is so ridiculous now, that I'm embarrassed I ever believed the tale in the first place.

### Plate Tectonics is Real

The problem is the same as that described in the earlier illustration with the eye-leg similarity. Like the other, the science community started with fact. Plate tectonics has been proven and evidence of its existence is seen every day. Then the scientific community went berserk because they couldn't figure out what caused these huge piles of rock. Instead of letting our children know that there are **"unknowns that should be explored"**, we continuously insulate, stagnate, and destroy the minds of our children.

Speaking of information, the diagram of our solar system that showed Mars and another planet needs to be explained before someone tries to put it into one of our Bode's Law kind of definitions. We may get insight from the ancient Zoroastrians. Here is what they had to say about the formation of the mountains.

*"As the evil spirit rushed in, the earth shook, and the substance of mountains was created in the earth. First, Mount Alburz **arose**; afterwards, the other ranges of mountains (kofaniha) of the middle of the earth . (arose).."* [These historians believed that some outside force created mountains around the middle of the Earth.]

## Computers to the Rescue

Thanks to many satellite pictures and computer models, we now know that something terribly bad happened to Earth and Mars and there is a high probability that they happened at or near the same time. According to computer models, Mars and Earth had several near collisions and we are going to discuss what happened when they came together. By the way, this is not some hair-brained concept that I pulled out of my head. Many reputable scientists are currently doing a lot of research in this area. These scientists first looked at the meteor evidence on Mars.

## Martian Crater Anomaly

Mars has a very unusual pattern of craters. **Ninety percent of all the cratered area is located on one side of the planet** and I bet no one ever told you that odd fact and even though you have probably seen photographs of Mars, it didn't seem strange until now. The crater side of Mars, as shown to the right below. Almost **all of the 2700 major craters that are over 20 "miles" across are located on this "bad" side.**

There are only two major ways that the strange cratering could have occurred. One way would be that a large planetary object got too close to Mars and exploded which, in turn, peppered the surface towards the explosion and left the other side unharmed. The mass would have had to be very close or the pieces would have been deposited around the whole planet as it rotated. The second way is similar to the first except that when the object got close to Mars, Mars itself split into two pieces, The cratered side we see today is what the whole planet looked like before the smooth half was ripped away. The second choice is the one that makes the most sense. Below on the left shows the side of Mars that has almost no craters. The crater side is shown on the right. Weird isn't it..

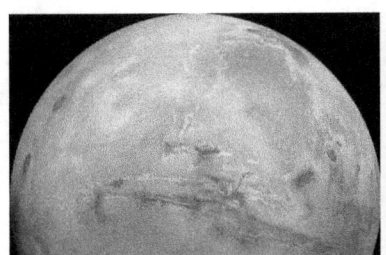 

### Close Encounter Model

This model confirms that Mars and the Earth came close together several times. In 1997 at Colorado's Laboratory for Space Physics, a team of scientists finally was able to model what researchers now believe to be the forming factor for our huge mountain ranges and the weird cratering on Mars. According to the evidence, Earth and Mars came close together on several occasions with disastrous effects. As the Earth went around the sun in the early days, its orbit was possibly more eccentric and it came very close to Mars at different times. Each time that it came close to the planet, it caused great perturbations on both planets.

### Pulled up Mountain Anomaly

On Earth, we can still examine the effect of the close flybys in the form of extremely long mountain ranges. Early theories that mountains were pushed up by plates moving together did not match the positions of the mountains and people began to wonder why the mountains all fell in straight-line patterns. One path is straight along the equatorial line and another goes along the side of the Americas and along the same path on the other side of the world.

### Another Plate Tectonic Error

Plate tectonic models should not go around more than 50 percent of the globe, because there would be no way to push the block, but the tectonic theorists had to make them that way to support the mountain ranges until sanity finally won out. New math models were able to capture the events that caused something quite different that mountains being pushed up. Instead, the mountain ranges were **pulled up**. A large planetary object strafing the planet made each of the extended mountain ranges. Once the Earth was strafed with the Earth rotating on an axis through the middle of the Pacific Ocean and Asia and a second time when the rotational axis was similar to our present rotational feature. Yes, I did say the earth's axis changed. In fact we will see that it has happened more than once. The picture below shows the projected path of the Mars close encounter on two separate fly-bys. The wide lines represent the long strings of mountains along the 2 paths. One going from the southern tip of South America through the tip of Alaska and down along the coast of the Far Eastern countries. The second, more severe uplift included the region from the Middle East through Pakistan, Tibet and China.

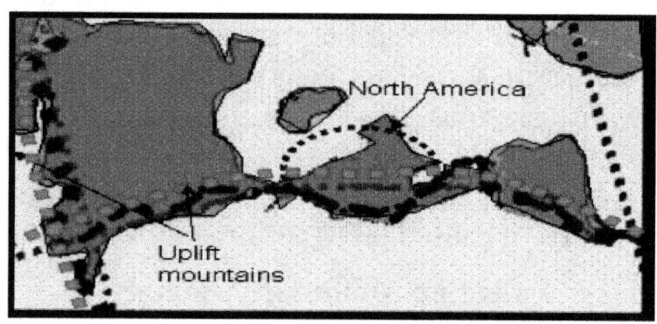

In the previous section we presented the ancient written testimony, which was tested during the Mars/Earth collision modeling and was shown to be accurate. The ancient texts told about the Earth being torn apart. You would think there would be some evidence of mountains rising high in the air during this time period and there is. Before the first "uplift" from the Mars encounter, the western side of what is now South America was underwater rather than being a huge mountain range. Today evidence of the top of the mountain being underwater can be readily seen. Next are piles and piles of HUGE prehistoric clams the size of a man. [ON TOP OF A MOUNTAIN IN PERU]

## Mars and the Pacific Ocean

During the last of these "almost" collisions, Mars got so close that part of what is now the Pacific Ocean was ripped away from the Earth. The time has been determined to be 400 thousand years ago. This would have been a third close encounter and we should talk about it here. Mars got way too close during the third flyby and a portion of the Earth in the area of the Pacific Ocean was ripped out of our planet. For us this was probably a good thing, because the Earth finally was pushed into a more stable orbit that was closer to the sun and no longer in danger of contact with Mars.

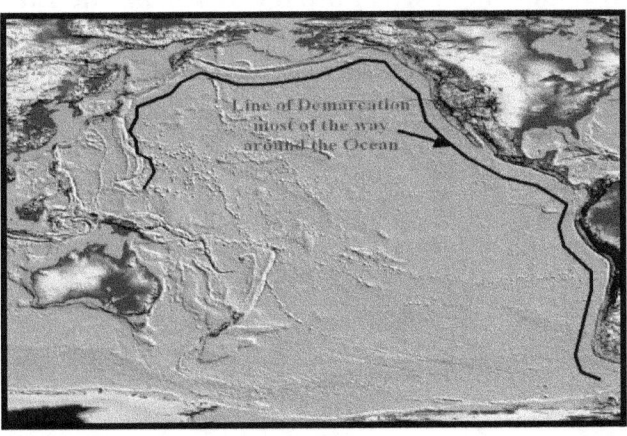

For Mars, the explosion left half of the planet scarred to this very day and the remaining portions of the debris are still in orbit around the Sun between Jupiter and the orbit of Mars. We know this remaining portion of the land that was located at the Pacific Ocean as "the Asteroids". Maybe we should call them the Pacific Asteroids. This event was captured in the ancient texts as the super planet was torn apart to make the earth and the sky. The Earth got formed and in the wake of its formation came great destruction. Somehow representative civilizations seemed to survive. We're going to discuss the cyclic destruction and reconstruction periods in the following texts.

## A scientific Version of the Creation of the post 400 Thousand year old Earth

Let's determine a "Scientific version of the Earth creation that is somewhat similar to those presented earlier from ancient times, but uses details of the mathematical and physical evidence collected to date.

- *A planet called [Earth] was in a widely eccentric path, that put it close to the planet Mars on occasion. **[Not the stable orbits you had believed.]***
- *On Earth and Mars lived many creatures. **[Oops! There goes that reference to ancient humans again.]***
- *During at least 2 of the flybys, the Earth's major mountain ranges were produced. [ **Not Plate tectonics**]*
- *Once, Mars came too close and a portion of the land, where the Pacific Ocean is, was pulled away from the Earth*
- *Many creatures were killed on both planets.*
- *The portion of the earth that was pulled away shattered and the pieces obtained an orbit around the sun to become the Asteroid belt.*
- *Many pieces hit Mars and caused great craters on only one side of the planet. The other half of the planet had been ripped away during the encounter. It is believed that some level of atmosphere still remained, but it was not to last.*
- *The major piece of the earth was cut loose from its former orbit and took a new orbit as a new Earth.*
- *One piece from the explosion stayed with the earth and became the moon.*
- *Violent storms and floods initially filled the site of the rupture and it became the Pacific Ocean.*
- *Pangea split apart to fill the rupture and the filling is still occurring today, as the Atlantic Ocean gets wider.*

## Pangea Anomaly

We have all been taught about how Pangea contained all the land mass of the earth and slowly separated <u>200 million years ago,</u> but there should be one nagging thing in your head. Why was all the landmass of the earth clumped together in the form of Pangea in the first place? With all the land mass in one tiny, tiny area, the roundness of the earth would have been questionable, but the size of the earth dictates that it will establish a spherical shape. The spherical shape would have had to have been established before Pangea was formed. If we believe that Pangea existed; and there is great amounts of data to confirm its existence; we are left with an unanswered anomaly. Certainly, our students would have been taught to question the whole Pangea concept on this one anomaly alone. So, we are left with the question. How could all the land in the whole world be clumped into one place? The answer is simple.

## It wasn't!!!!!!!!!

Before the big chunk of earth left, there must have been land generally around the earth to support the roundness requirement and the basic truth of Pangea's existence. This whole Pangea thing makes no sense unless there was an equally large mass on the other side of the world. Where did it go? I think you know and you should be getting pretty angry at the lie that has been told you? I know the historians and scientists have all lied to make our history more pleasant, but that isn't what is needed.

---

*The super continent of Prestonia was lost forever when the Pacific Ocean was made.*

---

# Mars Split In Two

Even with the above evidence many will say—Bah Humbug—to this whole concept. The near collision of Mars and Earth must be false because physical law requires that the smaller planet would have sustained the greatest damage. As we have already explained, THEY ARE RIGHT; at least in their theory. Let's look at the remains of Mars for a minute. I already mentioned that only half of the planets has significant amounts of cratering, but what I possibly didn't emphasize is the fact that while the earth split open during the flyby, Mars got the worst of it. It essentially split in two. That fact is so obvious, it is almost comical.

## Thin Crust

Just like the Pacific Ocean on earth with almost no crustal mass remaining after the split, over the ENTIRE northern hemisphere of Mars, the crust is rarely more than a few kilometers thick and the sparsely cratered surface is suggestive of a relatively new surface. Like the remaining portion of the earth, not including the Pacific Ocean, the southern hemisphere of Mars has a strikingly thick crust, which exceeds 20 kilometers in places, and a much more heavily cratered surface. It is in this hemisphere that we find nearly all the major impact basins such as Hellas, Isidis and Argyre with crater basins well over 1 thousand Kilometers in diameter. These huge holes were probably made by some of the large chunks of earth that left during the explosion. Mars also became a new planet, much smaller than it had once been.

## Evidence of a Split Planet

This odd cratering isn't most obvious evidence. Here is the proof from NASA. I know all of this sounds bizarre, so let's look at a topographical image of Mars. Please note that the northern hemisphere is not only smooth, but it also is sunken in much worse than our Pacific Ocean. It has a mean surface height 6 thousand meters lower than the mean of the southern hemisphere. Where do you suppose the northern half of the Planet went to?

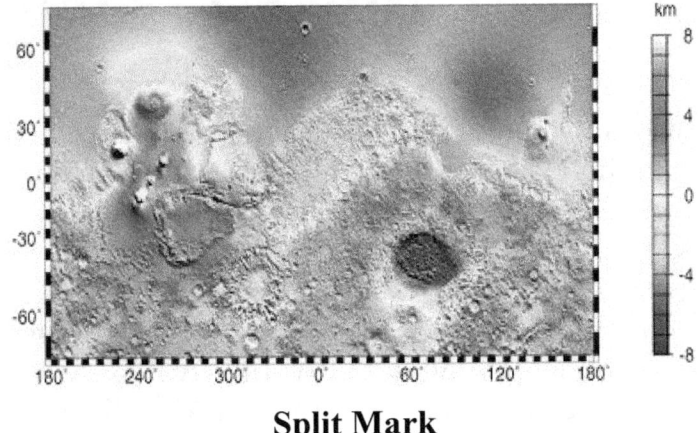

**Split Mark**

Also look at the dividing line between the hemispheres. Just like the huge slashes under the Pacific known as Mariana's trench, the northern hemisphere of Mars, is marked by one huge gash called the Valles Marineris. Although it is only 7 kilometers deep, it is up to 200 kilometers wide in places and has a total length of 4,500 kilometers. It is almost like Mars was split apart at one time and one of the marks of the split is this huge gash just like the Mariana's Trench in the Pacific.

**Healing Mars**

Mars, like the earth, is slowly trying to heal itself from this ancient event. In another 500 thousand years, the entire surface will probably be about the same height and here on earth, the Pacific Ocean will be about the size of the Atlantic.

# Dating the Mars Encounter

As I previously mentioned, the Mars incident occurred only about 200 million years ago. To test that answer we need to look at dinosaurs. The period from 200 million year ago until 65 million years ago is known for one thing—dinosaurs—BIG ones. At the end of this time, almost overnight, things got much smaller. There is a reasonable probability that one of the initiators for the large dinosaurs was the creation of the Pacific Ocean and the end of the age was quickly climaxed by the slowdown of the earth's rotation. I know those claims are not presented in other works, so let me start by listing things we think we know.

### The Earth Changed 400 Thousand Years Ago

If nothing else, the 400 thousand year old boundary is distinctive. Something happened to put a hole where the Pacific Ocean is today and this same event caused the super-continent of Pangea to begin to pull itself apart, and killed most of the living organisms on the planet. According to data collected from the Deep Sea Drilling Project [1968 to 1983] the **Pacific Ocean Basin** was determined to be youthful. This was determined by testing mantle depth and sedimentation over the ocean floor. The mantle is slowly repairing itself, but there is still a huge difference between the mantle thickness in the Pacific Ocean and that found on the rest of the world. According to many models, the **break-up of Pangea** occurred around 200 million years ago old dating system or 300 or 400 Thousand Years in the new dating. One simple test can be accomplished to test this date.

Major **meteors** hit at that time. So far, we have categorized at least 5 major craters that occurred on the Earth during that time. The craters can be found in France, two in Canada, the Ukraine, and North Dakota. Talk about destruction! That was probably the worst destruction ever. Over 90 percent of all creatures became extinct as the Earth entered the Triassic Period.

By looking at **huge piles of lava** in Siberia and the United States, scientists have dated major events in our earth history that happened around this same time. 400 Thousand years ago a 2 cubic kilometer pile of magma was formed in Siberia and another 1.5 million cubic kilometer pile of lava spewed from the earth at another location as it split open from meteoric or other external pressures. The date of the Pacific bottom, the date of the Pangea split, and the date of the most extensive extinctions point to the most traumatic event of our earth's history. That event was the encounter with Mars.

# Destructions

The list below is a shortened list of the near term extinction events that have occurred since this time.

| Years ago | Name | # of animals | % extinct |
|-----------|------|--------------|-----------|
| 400,000 | Permian | 80% | 95% |
| 350,000 | Triassic | 90% | 90% |
| 240,000 | Jurassic | 80% | 70% |
| 120,000 | Cretaceous | 90% | 85% |
| 11,000 | Venusian War | 100% | 33% |
| 10,000 | Pleistocene | 60% | 80% |
| 6,000 | Babel War | 33% | 33% |

One of the reasons for extinctions was Meteors. One of the meteors seems to have made the animals heavier.

## Lighter Animals

Have you ever wonder why the animals of the Jurassic, Triassic, and Cretaceous periods were so big? I'm sure you were told that evolution caused the gigantic monsters and then evolution caused them to die off. Well I'm sorry to disappoint you and say that that is rubbish. If evolution had been responsible for the change, then the change would have continued. There is a much simpler answer and for that answer we should first look at ice skaters again. Try to picture our ice skater spinning on an ice rink, if he wants to go faster, he simply makes himself smaller. He pull's himself into a tight space and he spins faster and faster. As he makes himself wider, he slows down. The earth did that very same thing 350 thousand years ago. When the earth lost the continent over the Pacific Ocean, there is little doubt that it began to spin faster. When I say little doubt, I mean the obvious. Animals got bigger. Even the insects got larger during this "special time". If the earth began to spin faster for a time, everything would become lighter and that is what makes things get bigger for no other "apparent" reason. Don't laugh at this theory, the evidence is pretty solid.

## Dinosaur Anomaly

If you have wondered why flying reptiles "supposedly" were not able to fly and why the dinosaurs "supposedly" could not run because their heart could not pump blood fast enough, the scientists reporting these findings did not help sooth your concern. The answer that they should have gravitated towards is that the flying dinosaurs **were** able to fly and dinosaurs were able to **run**. They were simply lighter than they would be if they lived today. If you are wondering why the oxygen levels are not as high today as they were during the Jurassic, the answer again is a faster spinning earth. Don't let some "artificial scientist" tell you that a flying dinosaur had to crawl up the side of a mountain so that it could open its wings and glide down to the ground hoping to snare some food on the way. I guarantee you that the species would not have survived. Don't let that same type of "scientist" tell you that a dinosaur could not run after his food. That animal would not survive. Don't let ANYONE tell you that a diplodocus could not lift his long extended head. If a dinosaur could not lift his head, he would be dead.

## Fast Earth Theory

After saying that things were lighter, one must try to understand how this very strange effect occurred so there is an important corollary I call the Fast Earth Theory. I know that the statement seems fanciful, but you can see its reasonableness if you look at Saturn, Neptune, and Uranus today. All of them are spinning so fast that things on their surface are much lighter than they should be. Even the atmosphere is lighter and all three are losing that valuable commodity known as atmosphere every day. This will continue until they slow down just as the earth did. This important theory provides a plausible reason why huge land dinosaurs don't exist today, but were able to exist in the past.

## Everything Got Heavier

The Martian event may have been a catalyst. When the earth got smaller and more dense, it began to spin faster and everything got lighter, but it did not sustain that spin rate.  It is apparent that about 50 thousand years after the near collision, the earth began slowing down its rotation rather over a short period of time. I know that's another wild idea that hasn't been  brought out by others so let me provide you with the evidence.

## Dinosaur Size Evidence

During the Jurassic, Triassic, and Cretaceous periods, the largest animals that ever walked on land were in abundance. The bone structures of these massive monsters don't exactly fit the animal structures that would be required today. If there is an anomaly it probably means that we haven't uncovered the truth. One such anomaly is shown with the Diplodocus. Its neck is too long. If the creature had tried to stretch out its neck in front of its body, the head would have come crashing to the ground as there is not enough muscle to pull it into the air unless its neck was lighter. The Tyrannosaurus Rex structure is such that it would not have been able to run in our atmosphere. I know some of the new theories have the Tyrannosaurus Rex as a scavenger, but that would only work if he could run. If it couldn't run it would have died.

Many of the Bird-like reptile would have surely died as they could not fly in our current atmosphere. To make this element more comical, scientists have invented many theories that suggest that the winged reptiles would have had to climb up to the top of a cliff and take off into the air. If they ever hit the ground they would be doomed as they could not regain flight. I know it sounds stupid, but that is the belief of many today. Hopefully you can appreciate the absurdity of such a theory. Still another problem is the dinosaur heart. It has been determined, by many, that the hearts on these massive beasts would not have supported the blood flows needed to sustain body function. Many have surmised that the dinosaur's blood flow problem could have been rectified by increased oxygen percentage in the air. The blood flow could be greatly reduced if more oxygen was present in each breath, but that does not account for many of the physical characteristic anomalies, and I'm not sure it would even satisfy the blood flow problem without other elements like lighter animals.

Scientists have proven that the oxygen content was higher during the time of the dinosaurs by simply sampling the oxygen captured in globs of amber. Therefore you have the question of why the oxygen was here during the Jurassic and not here now. I Think I can bring some sanity into this whole crazy mess.

*The answers are that animals were lighter in the old days because the earth quickly lost some of its size. It also had more oxygen in the atmosphere at the same time and lost much of it because it was spinning too fast to hold onto the oxygen for the same reason.*

## Less Density Evidence

If we make the assumption that the specific gravity was less during these ancient times, then there must have been a cause and we don't have to search long. Generally speaking, less gravity means smaller size, less density, or faster rotation. I don't believe that the earth got larger 120 thousand years ago, so the other two seem more likely. Concerning density, we have one very nice piece of evidence—the Moon. Below are three of the elements that we believe we know concerning the Moon and its relationship to the earth.

The Earth has a large iron core, but the moon does not. The Moon density is only about 50% of the earth's. This is because Earth's iron had already drained into the core by the time the moon was yanked away. The moon has exactly the same oxygen isotope composition as the Earth, whereas Mars rocks and meteorites from other parts of the solar system have different oxygen isotope compositions. This shows that the moon, most likely, formed from material in Earth's neighborhood. The moon orbit and Earth rotation are synchronized, suggesting that they both came from a common source.

I have presented an argument that the moon came from the earth, so let's just say that the Pacific Ocean opened up and spewed out the moon. The moon density level tells us that only the lighter portions of the earth were expelled, leaving the much heavier mantle intact. Therefore the specific gravity associated with the earth was much higher after the expulsion than before it and Dinosaurs would have been lighter before the Moon was made than equal sized animals from after this extreme event that reduced the earth's density. Today, the Moon is only 50% as dense and the Earth, so we can make a reasonable guess that the density of the earth was perhaps 10 to 15 percent less than it is today before that section was torn away. Therefore the dinosaurs would weigh as much as 10 to 15% less than they would today because of this factor alone, but that would not be enough difference. There must be another factor to be considered.

## Faster Rotation

The lower density wasn't the only thing that lightened the dinosaurs. We also have a pretty good example of what the earth might have been like with respect to rotation if we look at a picture of Saturn, below.

Saturn is spinning fast enough so that particles do not have to gain substantial amounts of energy to attain what is sometimes called escape velocity. Saturn's day is less than ½ of a "current" earth day and the whole planet is trying to escape. The bulge at the equator shows two things. It shows that components of the atmosphere are drifting away from the planet and it shows that gravity is lower. Remember the *"atmosphere drifting away concept for the topic below"*. What we find in our solar system is that, even though Saturn and Uranus are substantially larger than the earth, things on those planets are lighter than they are on earth. This is due to the high speed spin. On Saturn or Venus, animals would weigh only 90% of what they weigh on the earth and on Uranus they would only weigh 80% of their "Earth weight". One reason we can believe the earth was spinning fasted is the thicker air that has been lost over time.

## Thicker, Oxygenated Air

We can also surmise, from evidence described below, that before the earth slowed down, it probably had a thicker atmosphere rich in oxygen. The most logical way for the oxygen level to have slowly been reduced over time would be that our planet rotation was significantly faster than its present rate and similar to that of Saturn. Here is what researchers have found concerning the oxygen levels during the dinosaur days.

Experiments have been done to show that the oxygen content of the "Dinosaur's Air" was higher than our current air. This was easily accomplished by simply testing the oxygen content of captured "air" in ancient resins. The tests showed a substantial increase in the oxygen content when compared with "Modern Air". The problem with this experimental data is that the researchers only used the data to try to prove that a "water canopy" was around the earth before Noah's worldwide flood. Because the canopy theory had other problems, the issue was essentially disregarded. Because the air was thicker, the earth was most likely spinning faster to allow the oxygen to leave and other elements fall into place.

## Flying Reptiles that Flew

Remember the flying reptiles that couldn't fly? It would even be easier to fly and the flying reptiles wouldn't die if they touched the ground. Imagine that! Flying reptiles actually flew. Also the Archaeopteryx must have flown and its wings were not useless as you have been told.

## Why Would the Planet have Slowed Down?

There is a simple answer to this question. A huge meteor hit the earth and saved our planet, but the Dinosaurs were destroyed. They weren't destroyed by the K-T boundary that everyone talks about and the volcanic action or even the reduction in sunlight and reduction in plant life as a food source. They were destroyed by their own weight. I know there were heavy animals after the meteor strike of 120 thousand years ago, but generally their bulk had to be supported by water and the land creatures after that fateful time were limited to the size of an elephant. Even without Mars messing things up, the earth had its own internal problems. Its iron core was not and is not stable. Therefore, the rotational axis can "slip" from time to time. As it slips, hot spots form and explosions abound. Added to this is the continuous bombardment of meteors and comets along with the occasional exploding moon and you have an unstable earth.

# The Earth Goes Wild

The earth has been molded by many things besides Mars, so we should look at some of them to get a feeling about what the earth will do today and in our future. Some of the elements of effect are volcanoes, plate shifts, the axis of rotation flipping, and meteors. Each is described below.

## Volcanoes

While there is no evidence of the volcanic action needed to force mountains into the sky, that doesn't mean there weren't some big ones. Researchers have determined that there are peaks of volcanic action in earth's history that have occurred throughout time. From time to time the earth just blows off steam. Maybe tomorrow will be that day. Who knows. Our Earth is full of volcanic holes. Each one is ready to spread what is inside the Earth onto the outside. Some of the holes are spewing lava at this very moment. They will never make a major mountain range, but they can be very deadly.

## Taupo Volcano

The most ferocious volcano in recent history was on New Zealand. The Taupo Volcano erupted in 130AD shooting out 33 billion tons of pumice and debris. No other recent volcano has had that much pumice expelled. The ejected material covered a land area of 2,000 square miles. Since the Indian Ocean Tsunami in 2005, you probably know what happened next. The explosion caused a great tidal wave felt along the Australian coast. There were probably few survivors along the shoreline, but this was certainly insignificant when compared with major events in earth's history. From estimated dates of major eruptions one can tell that the earth goes through a higher than normal volcanic action time period about every 10 thousand years and the last one was about 10 thousand years ago. The earth is possibly ready to get into a higher volcanic action time period soon so don't go near active volcanoes if you can help it.

## Polar Axis Flips

This well tested theory provides striking evidence to support the "earth flipping on its rotational axis" phenomenon and its devastating effects. Haven't you wondered why they found thousands of animals quick frozen in Alaska and Siberia? The answer has nothing to do with some instantaneous Ice Age. It has to do with a warmer area quickly becoming the North Pole. Here's the scary part. According to Atlantic Ocean sea bed magnetics testing and piles of other evidence, this whole flipping thing happens a lot. One reason for Ice Ages is the erratic nature of the earth itself. You may not want to hear this, but the polar axis flips about every 100 thousand years and things get destroyed when it occurs. If you recall the graph I presented earlier, it showed the last 14 flips. Another flip could happen any time. Using mathematical models of the external crust and inner molten material, researchers have estimated with mathematic models that the Earth should flip on its axis about every 10 thousand years. The problem with trying to determine the actual workings of the Earth is that no one has ever seen the inside of the Earth to model it properly, but the results do confirm the high possibility of a polar flip, which will cause mass destruction, tidal waves, and major climatic changes. A polar flip, however, does not cause the most damage. Crust movement or magnetic field wander causes the real bad problems.

## Plate Shifts

Another way to look at magnetic wander is called plate shifts. Like the magnetic shifts, these apparent crust movements have been estimated to happen about every 10,000 years. One researcher indicated that the most recent ones occurred 43,000, 22,000, and 10,000 years ago. Sometimes the crust and magnetic field seems to wander over a number of years and other times it seems to jerk suddenly. In fact, the evidence suggests that there have been at least 170 major movements in the Earth's crust, which corresponds to the magnetic field shifts between the present day and the Cretaceous Age. One of the theories is that these "jerks in the crust are apparently caused by the uneven weight of the various plates supported on the surface of the Earth; especially the 19 quadrillion tons of mass called Antarctica which is located at the present day South Pole. Each time a movement occurs, terrible things happen like tropical areas turning into glaciers. Whether the evidence shows magnetic field wander or plate shift wander doesn't really matter, because the outcome is the same.

### Tropical Arctic

Researchers have found evidence that the Arctic was tropical for a short time around the end of the Cretaceous Period. As they investigated the depths, they found bones of early crocodiles, turtles and fish that were all tropical and estimated the summer temperatures reached into the 90s. The most logical explanation for the hot temperatures was that the plates shifted or the planet axis moved by a substantial amount many years ago. Finds similar to this have convinced many that the outer core of the Earth moves continually and that the movement is in jerks over time.

### Tropical Antarctic

If we move to the other side of the world, we find the same thing. The remains of tropical trees were found among the Magma beds from some ancient time. Additionally, they have found swamp type dinosaur bones along with remains of swamp type plants that existed before Antarctica became cold.

### Hot Spot Proof

As I brought out at the beginning of the book the Hawaiian Hot Spot is great proof of the erratic motion of the earth. For those thinking Hawaii is an anomaly, here are a couple more hot spot trails.

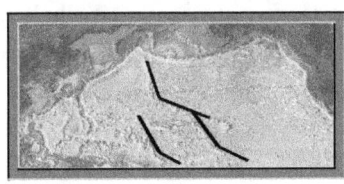

# Thousands of Meteors

Many of the extinction periods were brought on by elements outside our planet. One element is called a meteor. Many thousands of these meteors have hit our planet and scientists have built up a pretty good picture of major meteor strikes just by dating the craters they leave behind. The graph below shows 14 known major meteor hits, which caused unbelievably huge craters that are greater than 50 kilometers [30 miles] across. Imagine what would have made such a hole and imagine the destruction. There is more than a coincident timetable between extinction periods and the time of these meteor impacts and I don't mean just the one that we have all heard about that struck the Yucatan and signaled the end of the dinosaur age. I mean almost every destruction period was preceded by a huge meteor attack. The details following shows the immense size of some of these major impact craters. Some are over 300 kilometers [180 miles] in diameter. Whenever the meteors hit, a thin film of iridium dust sometimes would cover the surface of much of the Earth and allow scientists to categorize and time the event very accurately. Iridium is not found in abundance on the Earth except at these layers, so it's a really nice timing device. Below are 28 of the largest "known" impact craters found to date. Knowing that crater impacts on the water are much harder to find, you can image how many HUGE meteors have hit the earth.

# Major Cratering Around the World

| Where | Crater Dia. [KM] | When [Age] |
|---|---|---|
| Vredefort, South Africa | 300 | Precambrian |
| Sudbury Ontario, Canada | 250 | Camrian |
| Antarctica | 250 | Ordovician |
| Beaverhead Montana, U.S.A. | 60 | Silurian |
| Acraman, Australia | 90 | Silurian |
| Siljan, Sweden | 52 | Devonian |
| USA | 100 | Devonian |
| Australia, Chad, Africa | 125 | Pennsylvanian |
| Quebec, Canada | 54 | Pennsylvanian |
| Canada | 35 5 | Permian |
| Antarctica and Brazil | 50 | Triassic |
| Canada, France, Ukraine[5 craters] | 250 | Triassic |
| USA | 50 | Triassic |
| Puchezh-Katunki Russia | 80 | Jurassic |
| S. Africa, Norway, & Arctic Ocean | 70 | Jurassic |
| Queensland, Australia | 55 | Cretaceous |
| Yucatan, Mexico and Russia | 170 | Cretaceous |
| Nova Scotia, Canada | 45 | Tertiary |
| Virginia, Siberia | 100 | Tertiary |
| Kara-Kul, Tajikistan | 52 | Pleistocene |

Each of these huge meteors, not only had enough force to cause local extinction and massive cratering, but they also many had enough impact force to split open the Earth. Of course, I don't mean that the Earth was split in half many times, but there is strong evidence to indicate that the meteors caused great openings in the crust which allowed huge piles of magma to be expelled. That discussion is covered later, but just image a meteor that hits so hard that the earth begins to split open.

**Ending the Permian Age**-Let's look at the meteors that ended the Permian Age in particular. So far we have categorized at least 5 major craters that occurred from a massive planetoid that split up just before impact on the Earth during that time and "yes" the planetoid evidently was Mars as we looked at earlier. The impact craters can be found in France, two in Canada, the Ukraine, and North Dakota. Talk about destruction! That was probably the worst destruction ever. Over 90 percent of all creatures became extinct and that was just one of the times that an immense, impact crater, maker apparently caused mass extinction as we examine coincident timeframes.

**10 Thousand Years Ago-**While it paled in comparison to the Permian end event; a somewhat reasonably sized meteor evidently hit 10 thousand years ago and left a calling card over much of the earth's surface. The calling card was tektites. Tektites are small glassy bodies found strewn on and near the surface in several regions of the world. They come in various shapes: droplets, buttons, even dumbbells. By general agreement, tektites are attributed to meteoric or comet impacts that melt terrestrial rocks and splash liquid droplets into the atmosphere. Once in the air they are shaped by aerodynamic forces and solidify. This huge meteor hit somewhere on earth and strewed an immense batch of tektites and microtektites over fully 10% of our planet's surface, nearly 5 x $10^7$ square kilometers. Today this is called the "Australasian strewn field." Although the large quantity of these tektite things are found in Australia, the meteor hit in Indochina. In Indochina we find some of the largest tektites ever found [up to 24 kilograms over a 1,000 kilometers area. The vicious nature of the meteor strike would certainly have caused a level of extinction over a large portion of the earth's surface. It is not a surprise that the Pleistocene Era ended about this time with a massive extinction of animal life.

## Meteor Evidence

**The Chad crater**, which is over 100 kilometers across. The Meteorite storm that caused it also caused a major extinction during the Permian Age. Two chains of craters have actually been found in the area showing that whatever hit shattered into many pieces and struck many sites. To the left is the largest.

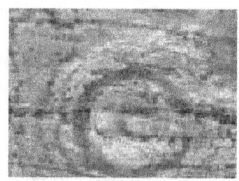 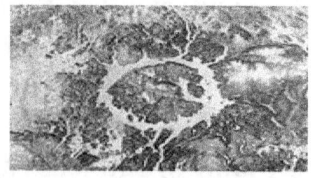

**The Canadian crater shown on the right** is over 100 kilometers across and is known to have been one of many that occurred at the end of the Triassic. It was part of the major group of meteorites that hit about the same time, as indicated previously.

**Another Canadian crater set-**The picture below shows two of the craters formed over during the Jurassic. One is 35km and the other is 25km in diameter. These could be remnants of the last great flyby of the planet Mars.

**The infamous Yucatan crater** occurred whenever the dinosaurs were destroyed at the end of the Cretaceous. It is over 170 kilometers across. This particular "hit" caused one of the iridium layers we use for dating and this layer extends around the entire Earth. The Yucatan meteor gave us a lot of iridium and took away those pesky dinosaurs. The only thing remaining is a perfectly round indention half on land and half in the Gulf of Mexico. The Yucatan crater is special in that large animals ceased to exist at the time it hit. As discussed earlier, the earth may have slowed its rotation and halted the loss of oxygen from the air at this time. Everyone got heavier and the loss of oxygen into the Solar System ceased. This level of stabilization spelled disaster for any huge animals.

**Antarctic Bone Bed-**The impact of this giant meteor left a calling card besides Iridium. Fossil deposits on Seymour Island, Antarctica tell some of the details. A giant bed of fish bones at least 50 square kilometers in area can be found there as if some sort of catastrophe had annihilated untold millions of fish. And you guessed it! This great bone bed was deposited directly on top of that layer of extraterrestrial iridium that marks the event.

### Where do Meteors Come From?

One of the most dangerous impact crater making strikes come from Comets. As the comets heat, large pieces of debris break away and form sort of a cloud around the huge mass of icy material that typifies a comet. The particles orbit the sun with the comet and when a comet comes near, watch out. Below are some of the more deadly comets. They are deadly in that they come closer to the earth than others. Sometimes comets even hit the Earth. As you can see, many of the comets come extremely close to the sun. Some get closer than the Earth is to the sun. Wirtanen and Schwassmann will be coming extremely close in the next few years. The year 2013 is of particular interest in that the Mayan calendar was abruptly halted at the end of 2012. Whether they knew something about this event is not known, but why they would have picked that date is surely a strange one.

## Close Encounters

Below is a chart of the main  meteors we are tracking-with name followed by the length of time to get around the sun, next encounter, distance [in AU] and magnitude comparison figure.

| Name | Yr | next | A.U. | Size |
|------|------|------|------|------|
| Halley | 76.1 | 1986 | 0.6 | 5.4 |
| Encke | 3.3 | 2003 | 1.3 | 9.3 |
| d'Arrest | 6.5 | 2005 | 1.3 | 8.3 |
| Tempel 1 | 5.5 | 2005 | 1.5 | 12.3 |
| Borrelly | 6.9 | 2001 | 1.4 | 11.4 |
| Giacobini-Zinner | 6.5 | 1998 | 1 | 9.3 |
| Grigg-Skjellerup | 5.1 | 1992 | 1 | 12.3 |
| Crommelin | 27.9 | 1984 | 0.7 | 12.3 |
| Honda-Mrkos | 5.3 | 1995 | 0.5 | 13.3 |
| Wirtanen | 5.5 | 2013 | 1 | 9.3 |
| Tempel-Tuttle | 32.9 | 1998 | 1 | 9.3 |
| Schwassmann | 5.4 | 2006 | 0.9 | 11.3 |
| Kohoutek | 6.4 | 1973 | 1.6 | 12.3 |
| West-Kohoutek | 6.5 | 2000 | 1.6 | 10.4 |
| Wild 2 | 6.4 | 2003 | 1.6 | 6.4 |
| Chiron | 50.7 | 1996 | 8.5 | |
| Wilson-Harrington | 4.3 | 2001 | 1 | 9.3 |
| Hale-Bopp | 4000 | 1997 | 0.9 | 1.3 |
| Hyakutake | 40000 | 1996 | 0.2 | |

## Sometimes The Earth Splits Open

The shifting earth causes many problems, but so has the occurrence of some of the largest pieces of meteorite material. The meteors sometimes hit so hard, the earth splits open.

**This is not a theory, by the way. It's a fact.** Many places around the world are nothing more than huge piles of magma that have surfaced after the earth split open. The Deccan area of India, for instance, is made up of <u>12 thousand cubic MILES</u> of LAVA piled up to make mountains that were pushed out at the end of the Cretaceous Period. There is no doubt about it. In the past, large meteors have hit so hard that the earth split open. This catastrophe has happened quite a few times and the evidence is astounding. Millions of cubic <u>MILES</u> of magma have spewed up in mountainous piles at the sites of the splits. The magma mountains can be seen around the world and, guess what!; scientists have known about them for a long time, but no one teaches the scary facts. Critics say that meteors alone would not cause the kinds of extinctions that have been witnessed over time and they are absolutely right. In addition to huge meteor craters, many times the Earth was actually split open by meter strikes or similar action. I know that sounds scary, but evidence is evidence. These splits typically caused **millions of cubic kilometers** of lava and debris to spew out. The smoke and debris from this action sometimes covered the Earth with material that could block out the sun. What I think is the most enlightening part is that the split usually occurred on the opposite side of the Earth from the meteor hit. Just think about how much force would be required to do that. Here are some of the major lava flow events that have occurred since the Jurassic Period. The events I'm talking about here are not the occasional volcanic eruption. These are the effects of the Earth splitting apart in a huge area and spewing out **millions** of cubic kilometers of lava. The graph shows some of the major events that caused lava flows in excess of 500 thousand cubic kilometers.

| | |
|---|---|
| **During the Triassic** | Siberia [2 million cubic KM] |
| **During the Triassic** | USA [1.5 million cubic KM] |
| **During the Jurassic** | Karoo, Africa [2 million cubic KM] |
| **End of the Jurassic** | Serr Gal [1 million cubic KM] |
| **End of Cretaceous** | Deccan, India [2 million cubic KM] |
| **End of the Tertiary** | Ethiopia [1 million cubic KM] |
| **In the Pleistocene** | Ross Sea, Antarctica 1 million cu. kM |

The picture below shows one of the largest piles of lava. It is in India, in fact it used to cover just about the whole country and it still covers over 200,000 square miles and is 6,500 feet thick. It is estimated that originally it covered 1.5 million square miles and contained over 12,000 cubic MILES of lava. [**That's right I said MILES, not meters**] When the Earth splits open it splits wide open.

As I mentioned before, one of the times that it split open, a huge chunk actually split away from the Earth to make the moon. The Pacific Ocean was also created at the same time. By the way, even when these huge lava eruptions occurred, no new mountain ranges were formed except by the piles of lava. And if that force wasn't enough to push two of the plates fast enough to cause the Himalayas just what was supposed to have caused the great plate smashes? Schoolbooks typically don't cover these huge lava flows because someone might ask that very question. They also don't talk about the Earth splitting open because someone might get uncomfortable when thinking that the Earth could possibly come apart. Unfortunately, it could and being comfortable doesn't change reality, my friend.

### Are Meteors Harmful?

Some believe that, except for very rare occasions, meteors don't really hurt anything. What about this list below? This is just a small sampling of the many disasters caused by meteorites in recent history. These things fall all the time and many times people are killed.

# Meteoric Damage

| When | Where | Reported effect from Meteors |
|------|-------|------------------------------|
| 616AD | China | 10 deaths from meteor shower |
| ~1350 | China | Iron rain killed people & animals |
| 1490 | Shansi, China | Stones fell over 10,000 killed |
| 1511 | Lombardy, Italy | A Monk killed |
| 1639 | China | Large stone fell, 10s killed |
| ~1640 | Indian Ocean | Two sailors killed |
| ~1650 | Milano, Italy | Monk killed |
| 10/30/1801 | Suffolf, England | Home set on fire |
| 1/16/1825 | Oriang, India | Man killed; woman injured |
| 12/11/1836 | Macao. Brazil | Several homes damaged |
| 5/1/1860 | New Concord, Ohio | Horse killed |
| 6/30/1874 | Chin-kuei, China | Child killed, house crushed |
| 1/31/1879 | Dun, France | Farmer killed |
| 3/11/1897 | Martinsville,W. Va. | Horse killed and man injured |
| 6/28/1911 | Nakhla, Egypt | Dog killed |
| 9/5/1907 | Hsin-p'ai, China | Whole family crushed to death |
| 6/30/1908 | Tunguska, Siberia | 2 killed, many injured |
| 7/19/1912 | Holbrook, Arizona | 14 thousand stones fell |
| 12/8/1929 | Zvezvan, Yugoslavia | One killed in a bridal party |
| 11/28/1954 | Sylacayga, Alabama | Woman hit by 4-kg meteor |
| 8/14/1992 | Mbole, Uganda | 48 stones fell, one boy hit |

## Can It Happen Again?

From the thousands of planetoids of significant size that are being tracked, beyond those classified as comets, we can make another short list of the ones we expect to come close to us in the near future. Chances are, something will hit us in the near future. It might be 10 years or it might be a thousand years, but we will feel the assault of these intruders.

On the next page is a list of those which will pass within 900 thousand miles of our planet, unfortunately the accuracy of the tracking is only about +/- 900 thousand miles until the objects get relatively close to us, so who knows. Note also that some come close to us often. The Planetoid named WN5 is one such nemesis which has a "near" collision scheduled in 2039.

## Collision Asteroids

| Name | When | How close [AU] & [MILES] | |
|---|---|---|---|
| AD2 | 2133 Apr. | 0.0087 | 800,000 |
| AN10 | 2027 Aug. | 0.0027 | 250,000 |
| CU11 | 2080 Aug. | 0.0043 | 400,000 |
| DB7 | 2048 Feb. | 0.0079 | 700,000 |
| DU3 | 2143 Mar. | 0.0067 | 600,000 |
| DV9 | 2160 Jan. | 0.0066 | 600,000 |
| Hathor | 2069 Oct. | 0.0066 | 600,000 |
| Hathor | 2086 Oct. | 0.0060 | 500,000 |
| HH49 | 2023 Oct. | 0.0079 | 700,000 |
| LV | 2076 Aug. | 0.0071 | 600,000 |
| MN | 2010 June | 0.0076 | 700,000 |
| Nereus | 2060 Feb. | 0.0080 | 750,000 |
| Nereus | 2166 Feb. | 0.0057 | 500,000 |
| NN4 | 2144 June | 0.0048 | 450,000 |
| NY40 | 2002 Aug. | 0.0035 | 300,000 |
| OX4 | 2148 Jan. | 0.0020 | 180,000 |
| QK130 | 2128 Mar. | 0.0097 | 900,000 |
| RQ36 | 2060 Sept. | 0.0055 | 500,000 |
| TU28 | 2121 Apr. | 0.0065 | 600,000 |
| TU28 | 2102 Apr. | 0.0079 | 700,000 |
| UG11 | 2008 Nov. | 0.0091 | 800,000 |
| VP11 | 2086 Oct. | 0.0060 | 550,000 |
| **WN5** | **2039 June** | **0.0015** | **125,000** |
| WN5 | 2028 June | 0.0044 | 400,000 |
| WO107 | 2093 Nov. | 0.0079 | 700,000 |
| WO107 | 2140 Dec. | 0.0005 | 50,000 |
| XF11 | 2028 Oct. | 0.0062 | 550,000 |
| YB5 | 2002 Jan. | 0.0056 | 500,000 |

The possibility that a huge comet or meteor will strike us in the very near future is compounded by visions provided to us from the famous 15th century "seers" Mother Shipton, and Nostradamus. These modern seers and others have projected the outcome of a very near term disaster. I will discuss the probable future comet or Meteor strike later. Before that terrible time, some precursor meteors have fallen.

## Our Last Major Meteorite Strike

No one seems to talk about this event, and it happened in a very remote section of the world, but it could have been very noticeable if it hit anywhere else in the world. The date was **December 9, 1997.** At 5:11 A.M., crews of three trawlers at widely separated sites off south Greenland reported "a blazing fireball that turned night into day." At a distance of over 60 miles away, the flash was compared to that from an atmospheric nuclear explosion. Seismic tremors also emanated from Greenland, so the impact of a large meteorite is almost certain. So far, no one has found the remains of the huge meteorite, but you have to recognize how very desolate and impossible that area is to search. Luckily our last major meteor hit Greenland and not Disneyland or people would easily accept the event as a real danger. The 1997 event was small in comparison to those that will happen in the near future. One of the most likely, near term, candidates is about 2 Kilometers wide and it is classified as NT7.

## Eminent Asteroid Strike

One asteroid not on the list is one named NT7 was just discovered in 2002 and is on an impact course with Earth. If you thought the others were scary, this one is expected to strike or come extremely close to our planet on 1 February, 2019 and even though it is only 2 kilometers wide, it will be traveling at a rate of 28 thousand miles per hour and contains enough energy to cause continent-wide devastation on Earth, so we should not ignore this terrible threat. To make things worse, if it doesn't hit then, this rock circles the sun every 837 days, so it will have another shot later, but it's not the only chunk of rock that is probably heading for us. To make things look really bad scientists now track a huge quantity of these asteroidic masses. The picture below shows those that are close to the earth. Each dot represents a tracked asteroid. The circles are the orbits of the major planets close to the sun and a portion of the orbit of Mars is shown as the outside circle. What I really wanted to show here is that the probability of getting hit is higher than most want to believe.

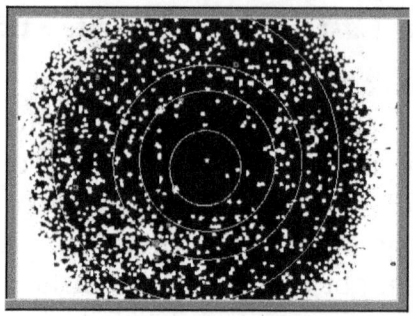

Besides tracked data, we can use two things that are a little more exotic to find out about upcoming meteoric disasters. The first thing is to know that history repeats itself. Many meteors and even comets have hit in the past and they will hit again. Although tried and true, that method has no timing accuracy or reasoning. The second method is by listening to prophets or seers, but most who claim to foretell the future do not. Some special people seemed to have been able to foretell the future and we will need to investigate further.

If one cannot see the future, one should, at least, be able to understand the past. One thing to understand is that while all of the catastrophes on earth were happening, animals flourished. The expansion of the animal kingdom was way too fast and furious. Some say animals got here by some random "survival of the fittest " mutation redone millions of times over millions of years, some say God created everything in 6 days, some say the animals were seeded by aliens from another world. I say let's look at the evidence. First we will investigate the misinterpreted Creationist Theory of Evolution by God's direct creation in 6 days a little more.

# Ice Ages

Another resulting effect of crustal movements and environmental stress is the Ice Age phenomenon, which destroys creatures and molds the planet. In the very distant past, this phenomenon had very limited scope, but something very strange has been happening over the past 400 thousand years.

*Prior to 400,000 years ago Earth did not appear to go thought the rapid thermal changes that have been witnessed during the Pleistocene Age and no one knows why.*

All of a sudden 13 major changes occurred and scientists tracked the changes by checking the ratios of oxygen isotopes in Planktonic foraminifera [tiny dead animals] brought up from deep-water sea cores. The ratio tells the temperature when the animals died and Paleo-magnetic dating the cores tells the date. The table shows what was generally found in the North Atlantic Ocean. Water height data was added from sedimented shoreline fossilization and remains.

Current dating places the beginnings of the recent ice ages at 48,000, 33,000, 22,000, 18,000, and 13,000 years ago.

The graph following shows how the water height variations may has shown up in the mid-Atlantic. The portion from about 10 thousand years ago to the present does not come from the north Atlantic wave height data, but instead is derived from historical data and written testimony.

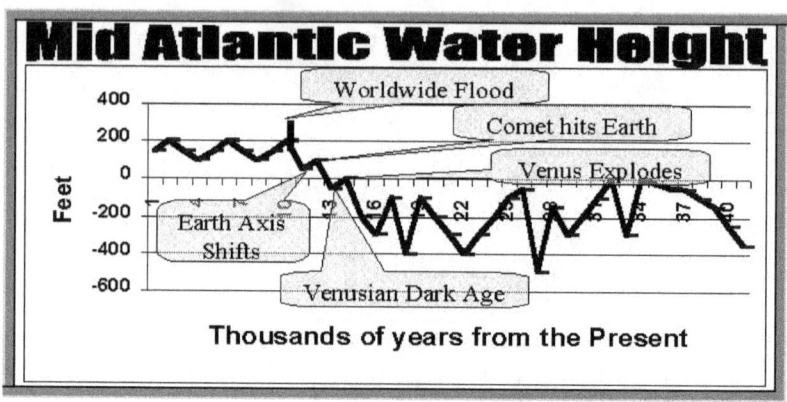

**Mid Atlantic Water Height**

Worldwide Flood

Comet hits Earth

Venus Explodes

Earth Axis Shifts

Venusian Dark Age

**Thousands of years from the Present**

You will note from the table that about 10 thousand years ago the water level increased by 200 feet after increasing about 200 feet only a thousand years earlier. All this water level increasing happened around the time that we hear stories about a place called Atlantis and at least 6 other major civilizations that were engulfed in water and a worldwide flood that followed a few thousand years later. The drawing below shows what the world looked like with the water level 500 feet below what it now is. One thing that pops up is that the Azores becomes a huge island in the middle of the Atlantic Ocean, the Red Sea became a river, there is a huge island in the Indian Ocean, and the Mediterranean was simply a group of huge lakes connected by rivers.

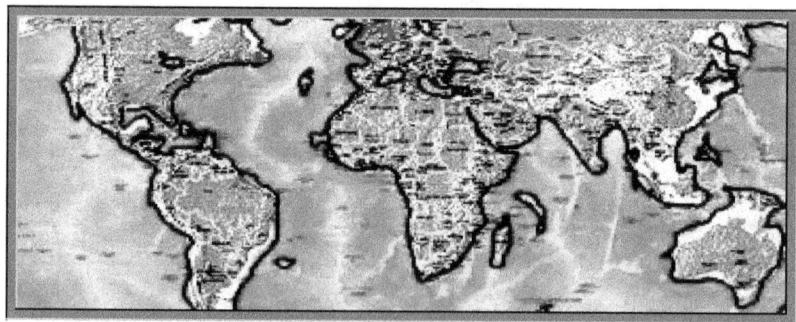

# After the Split

## *Destruction Regularity*

Don't worry if you disagree with the dates of these elements. That is your prerogative, but at least stay open to the possibilities as you find evidence in this work and other similar historical works of the following. The earth shifts, gets bombarded, splits open, and changes its rotational speed.

*The earth is not stable. Its planetary neighbors are not stable either.*

## Entry Into the Modern Period

Now that the earth settled down at its slower rate. Everything was ok of a long time until Venus got in the way. Before we get into the next time period. We need to step away and look at the other planets. When we get to the discussions about Venus, the 11 thousand year old destruction period followed by massive shift in the earth and worldwide flooding marking the beginning of the Modern Period will make sense.

Scientists call it the Holocene Boundary. Evidence shows that this period is marked by the earth flipping on its axis and causing Mammoths and other animals to be quickly frozen in their tracks. Some have been found with food still in their mouths. The rotational axis of the earth shifted about 30 degrees during this time. Places that were cold became less cold and hot areas began to freeze. I guess you can imagine what happened. The polar Ice caps melted according to the evidence and many areas flooded. As many commercial centers were located on Islands, commerce declined drastically and stories were told of some of the sights. One was called Atlantis by some, but there were others.

While we don't know exactly what caused this period of instability, you can tell from the previous information that it wasn't all that uncommon. One of the instigators was Venus. From a horrible event in the evolution of our planets, the moon of Venus was destroyed. The explosion rocked earth. Hundreds of meteorites fell along the equatorial path and ancient Jewish historical works along with others tell us that about 1/3 of all life was destroyed during the turmoil. Before we get into the details of the planet changing union of Venus and the Earth, let's look at the other planets during the time preceding this critical event. While looking at all the planets and large solar system bodies would be unfruitful, I will limit the discussions to the planets that could have sustained life in the old days.

# Our Solar System

Just like the unbelievable stresses seen on earth, the planets of the solar system went through many "changes". Before we can decifer planetary details we first have to start by knowing what our solar system is. What does our Solar System look like? Today, most will say that there are 9 planets and some will say Pluto is the farthest from the sun while others will correctly say that Neptune is farther away from the sun than Pluto, but what we don't generally consider are the asteroid belt, the Oort cloud, the Kuiper Belt, and several other large things going around our sun. Each system is very important and should not be ignored.

We found out that the Asteroid belt is, more than likely, the debris from a near collision between Mars and the planet that eventually became the Earth, but what of the other things mentioned? The kuiper belt contains at least 31 planitoid bodies including the fairly well known object we call Pluto. Whether objects in this area are planets or not have been debated for years including the idea of Pluto's planet status. Immediately outside this string of planetoids is a spherical cloud of various size masses that have even more eliptical orbits than Pluto. Most of these planetoids are called Comets as they show off a tale as there orbits comes closer to the sun. Some believe that very large masses in this area which extends out 50 thousand times as far from the sun as our Earth, about 5 trillion miles, could contain some form of life. I'm not going to get into that, but during the ancient times, people wrote about living on other planets.

## Haggadah

As part of the Talmud, the ancient Jewish book of "Haggadah" tells us about inhabited planets. While the book itself was put to print around 2nd century BC, it was made up of word of mouth

113

stories kept for centuries as the only way to effectively keep a historical record. The book says, "Nor is this world inhabited by man the first of the things earthly created by God He made several worlds before ours, but he destroyed them all." It goes on to explain something about some of these inhabited planets beyond earth.

- " *Erez, the lowest [7$^{th}$] earth lie in a succession the abbyss, the Tohu, the Bobu, the sea, and the waters.*"
- "*Adamah is the 6$^{th}$ earth. This is the scene of magnificence.*
- "*Arka is the 5$^{th}$ earth. It contains Gehenna, and Sha'are Mawet, and Sha'are Zalmawet, and Beer Shahat, and Tit Ha-Yawen, and Abaddan, and Sheol*"
- "*Harabah is the 4$^{th}$ earth. It is the place of brooks and streams*"
- "*Yabbashah is the 3rd earth. It is the place of rivers and springs*"
- "*Tebal, the 2$^{nd}$ earth, is the first mainland inhabited by living creatures, 365 species, all essentially different from those on our own earth. Some have heads set on the body of a lion, or a serpent, or an ox, or have other human bodies topped by the head of one of these animals. Besides, Tebel is inhabited by human beings with 2 heads and 4 hands and feet, in fact there ar organs duplicated excepting only the trunk. Sometimes these double persons quarrel with themselves. This species is distinguished with great piety*

**The first is our own earth.**-Whether this idea of inhabited planets is a possibility or not, our children should not be confined to think of the Solar system as a group of 9 planets would suggest and they should not be halted from thinking that there possibly was life on our close celestial relatives. Let's look at the inhabitable planets.

# Livable Planets

This is not about clusters of beings that spontaneously erupted on different planetoids throughout the galaxy. The probability for that is astronomical. What is more likely is that the ancient humans; those living over a hundred million years ago, finally advanced to the point where space travel was an obvious necessity and was commonplace. The reason that discussion of visitation is important is that sometimes the affect was disastrous to the planetoid itself. In fact, planetary visitation or the stresses caused to secure colonies might have been the most stressful element on some of our close neighbors.

Before I get into the most obvious sites for space travelers in the ancient times and the destruction left behind, let's look at some of the less obvious places that may been inhabited years ago. It may be surprising to some that we have so much data from places so far away. Below is a list of the "suspect planetoids" and there sizes. The planetoids that are shaded are those less likely to have ever contained human life. Possibly the important thing in this list is that many of the planetoids are not shaded and possibly had outposts located on them.

| Rank | Name | Dia[M] | Rank | Name | Dia [M] |
|------|------|--------|------|------|---------|
| 1 | Jupiter | 71492 | 9 | Titan [S] | 2575 |
| 2 | Saturn | 60268 | 10 | Mercury | 2438 |
| 3 | Uranus | 25559 | 11 | Callisto [J] | 2400 |
| 4 | Neptune | 24764 | 12 | Pluto | 2200 |
| 5 | Earth | 6378 | 13 | Io [J] | 1815 |
| 6 | Venus | 6052 | 14 | Moon [E] | 1738 |
| 7 | Mars | 3398 | 15 | Europa [J] | 1569 |
| 8 | Ganymede | 2631 | 16 | Triton [N] | 1353 |

## Beyond Pluto

The above list does not make the total sum of major planetoids. Even though you are not typically told about two more beyond

Pluto, they have been studied over the last 15 to 20 years and have become our closest known major planetoids beyond the orbit of Pluto. One is named Veruna and the other is simply named KX76. Planetoid KX76 has been tracked now for the past 18 years. It has been determined to be only slightly smaller than Pluto at 1.5 km in diameter and its orbit is very similar in its elliptic shape, but its average orbit distance is 6.5 billion km away from the sun while Pluto orbits at about 6 billion km and has a diameter of 2.2km. Pluto also has a moon named Chiron that is 1.2km in diameter, so it becomes the 3 largest planetoid in the Kuiper belt of our solar system. Planetoid Varuna is the next largest planetoid with a diameter of about 1 km. The obits of these somewhat minor planets is shown below, but if we look even past the Kuiper belt and into the huge pile of dust captured by our sun called the Oort cloud, we may find an even more spectacular find that is typically not brought out.

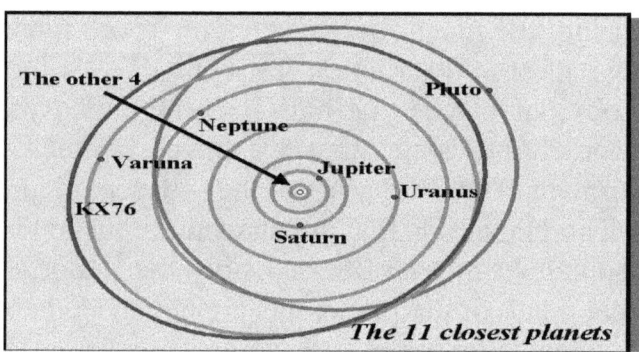

*The 11 closest planets*

There is no evidence to suggest that any of the extremely remote areas ever contained humans, but some believe in a planetoid called Niberu.

### Possible Niberu

Zecharia Sitchin interpreted the ancient Sumerian texts to read that the seed planet of Earth was a giant planet, Niberu, well beyond the orbit of Pluto; in fact, it supposedly takes 3600 years just to circle around the sun. Scientists haven't found that particular planet yet, but a huge planet seems to be going around the Sun that is a monster. It's being called the backward planet. This planet could possibly be like Niberu, but it is so far away that information detailing the monstrous planetoid is slow in coming and interpretation is tedious.

**The Backwards Tenth Planet-**In 2002 a UK astronomer may have put together enough pieces to have officially discovered a new and bizarre planet orbiting the Sun, 1,000 times further away than the most distant known planet. This NEW planetary body would be 30,000 times more distant from the Sun than the Earth. Not much is known about the object, but Dr. John Murray, of the UK's Open University analyzed the orbits of 13 comets and has detected the tell-tale signs of a single massive object that deflected all of them into their current orbits and he has estimated the size of the planet to be several times bigger than Jupiter. Professor John Matese, of the University of Louisiana, has determined the same thing in completely separate studies. The picture above shows the relative sizes of the planets to this new found monster. Being so far from the Sun, three trillion miles, it would take almost six million years to complete an orbit and it does the whole trip backwards. That's right, its orbit is opposite all of the other planets. From big let's go small to the closest planet in our solar system.

## Mercury

Mercury is a strange planet in that part of the planet is too cold to live on [40 Kelvin or minus 230 degrees C.] while half of the planet is always blistery hot [700 Kelvin or plus 430 degrees C.] Whoever wanted to stay on that planet would have to find an in-between spot and he would have to move his site every day or so as the rotation of the planet is about 1.1 times the revolution around the sun. Theoretically, it could be done. So now let's look for water. Some believe that it is too close to the sun to have been there very long. If that is so, it might have been captured fairly recently. Another reason the capture possibility has gained acceptance is that the rotation and revolution are almost linked, just like the linked orbit of the moon around earth. This link suggests that mercury came from the sun itself just like our moon came from the earth as we previously discussed.

**Mercurian Water-**<u>On Mercury they found water</u> on both poles. More water is located on Mercury than there is on the Moon. Because its rotation and revolution cycles are so similar that there could have once been life. At the poles there could have been some semblance of life provided that the population slowly migrated to insure that they stayed towards the "dark" side of the planet and along the pole. While Mercury has a very slight chance that humans had ever visited this planet, Mars and Venus provide us with very good indications that humans once roamed on those planets. I'm skipping over Venus for a minute and going right to Mars and several others. Venus has affected the earth's evolution in the very recent past, so that will be studied later. Right now let's travel to Mars.

# Mars

Before we get into discussions about the formation of Mars as it is today, we must first look at things that might be considered anomalous.

- *Mars is much less massive than any other planet, except Mercury and Pluto which have been generally considered to have been escaped moons.*
- *The Martian southern hemisphere is peppered with craters. Few are seen in the north.*
- *The crustal dichotomy is almost a perfect circle. This suggests that an extremely close cataclysm.*
- *The smooth crust of the northern hemisphere is only about a kilometer thick, compared to 20 kilometers in the south. This indicates that the northern half of the planet has been ripped away. While this fact is unmistakable, some choose to ignore it.*
- *The nearby asteroid belt is clearly the remains of a planetary mass that once was spinning around the sun. The problem is there isn't nearly enough mass to have been an entire planet.*
- *The huge erosion patterns on the surface indicate massive flooding in the past. The very recent past.*
- *Many geometrical shapes with multiple parallel and perpendicular lines can be found. If these were found on earth, people would indicate that they were the remains of an ancient city or cities..*
- *Long straight lines that go one for miles just like roadways.*
- *Areas that appear to be burned. The areas look like burned out cities in the aftermath of a violent war.*
- *Reduced Iodine-129 [Xenon-129] can be found in abundance indicating some type of "nuclear events or massive*

*explosions" were close to the Martian surface. That sounds like war.*

In addition to the anomalous physical features and thanks to many satellite pictures, we can be fairly certain that life was on Mars many thousands of years ago. We also can be fairly certain that people of earth greatly affected the Martian development. This thought does not violate scientific reason. It simply means that there was a very ancient group of humans that we know very little about. This ancient group of humans probably went everywhere including Mars. Because people are the way they are. There soon were many wars including wars between planetary colonies. Much of the destruction found on Mars did not come from it splitting apart. It also did not come from the planet slow loss of atmosphere and water.

## Mars had a Substantial Atmosphere

It should be noted here that there are planetoids in our solar system besides earth that contain oxygen in the atmosphere. Those with the most and best were Mars, Venus, and Europa, one of the moons of Jupiter. Sometime in the past; Mars began to lose its "normal oxygen" atmosphere. Finally, all the inhabitants left or were destroyed. We know that Mars had a substantial atmosphere in the recent past because the surface of Mars has evidence of substantially fewer craters than found on the moon and the ruins of many buildings are still visible today. That doesn't mean that there are no craters of Mars, in fact, there are about 2700 that are over 20 miles in diameter and 15 that are more than 130 miles in diameter, but **almost all of them are on one hemisphere of the planet** [the southern hemisphere]. The percentage of craters to landmass on Mars and Venus are both substantially lower than that found on the moon because of one thing. Recently they both had substantial atmospheres. Venus still has its atmosphere, but the Martian atmosphere is almost completely gone. Here is the probable truth. Before some terrible changes occurred, Mars was inhabitable by humans and there is evidence still visible that this inhabitable place was inhabited.

Timing events on Mars is extremely difficult but the idea that humans lived there removes some of the anomalousness of the evidence that can be found. The most likely reason areas look like cities is that they once were cities. The reason that these areas looked like they were burned out, most likely is that they were, and the reason the buildings look similar to earth buildings is that earthlings most likely built them. Humans molded Mars and, most likely greed or someone trying to increase his "power" caught Mars unaware and soon it was too late. The atmosphere, the rivers, the lush vegetation, the laughing of children all disappeared.

## Martian Cities

From photographic evidence and from research initiated by researcher Richard Hoagland, the remains of human existence can be found everywhere on Mars. Much evidence is covered in years of dust, but is still recognizable. The picture below and associated drawing shows the remains of only one of the seven "cities" that have been "found" on Mars by the various picture taking probes that have been sent in the recent past.

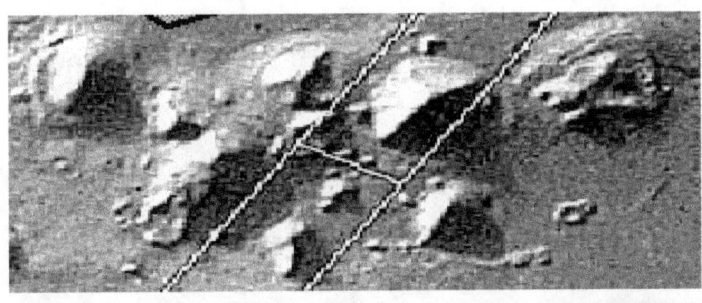

The drawing tries to show what the city might look like with the major streets uncovered. Even with a substantial amount of damage to the cities, there still is a significant amount of detail including <u>right angle walls</u>, pyramid structures, and walled courtyards Note the almost perfect, huge pyramid [second building from the right].

Let me say that my pictures do not do justice to the actual finding of the various researchers on this topic. My desire here is to let you know about the findings and provide an overview. Not only does the above "city-site" seem manmade, but other areas have possibilities of showing signs of life as we send more probes to our closest neighbor. More details concerning Martian habitation can be readily found through details collected by many sources. The premier investigator is a man named Richard Hoagland but there are hundreds of people who have looked over the photographed areas of Mars and other planets and have reported on what hey saw. Don't just look at this brief overview and think that this is the only evidence of human existence on Mars.

I am only presenting a small quantity of elements just to show that there is enough evidence to even allow the most ardent "non-planetary habitationalist" cause to wonder. The photograph below shows the possible city site and the now familiar area known as the face on Mars. Whether or not the apparent face is a "manmade object or not is not known by the author, but it is interesting to note how close it is to town.

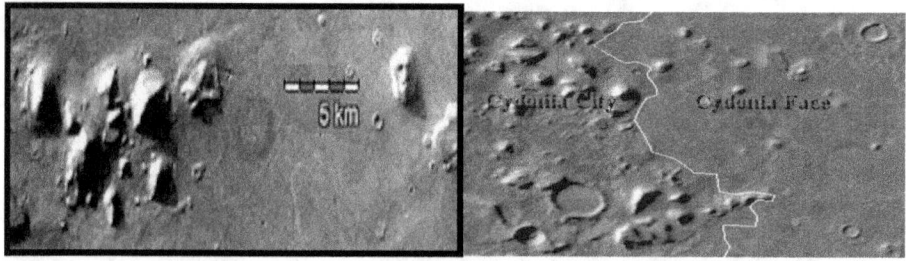

Speaking of peculiar, what researchers have found is that this "city" is right on the border between to smooth side and the cratered side of Mars as shown in the map above right.

## City Number Two

It is improbable that a city would have survived the moving of the mantle on Mars after the near collision with earth. Therefore, the most likely scenario is that this dwelling area and the others nearby were all built and inhabited  well after the 212 million year old date. The second "city" shown below contains many well defined pyramidal "buildings" that are positioned in a matrix that resembles a city. The pyramid in the upper right is especially interesting in that, even the steps up to the building can be made out. The drawing below may bring out some of the features. By the way there are better pictures of this area and the other areas shown. They can be seen in books by other researchers. My mission here is to simply bring up the probability. I placed lines on the picture to signify major roadways.

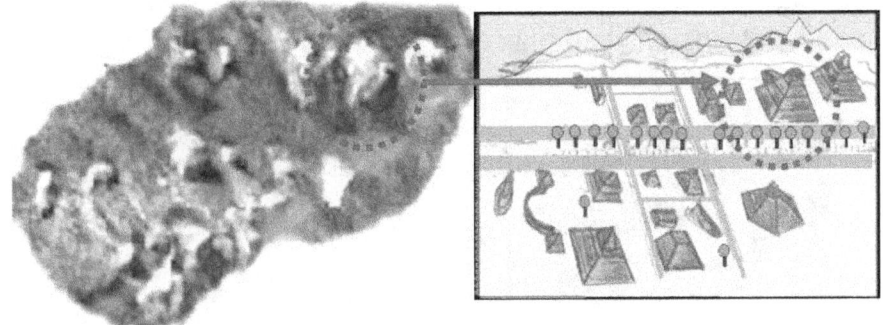

## City Number Three

The  third "city" following shows the effects of some terrible ripping on Mars as a huge crevice splits one of the cities in two. Also a crater can be seen in the city itself with a large building almost to its rim, showing that the city may be older than the crater. Right angles can be made out everywhere in the city area, which is not typical of non-manmade structures. Again my pictures are poor, but look especially at the shadows to examine details.

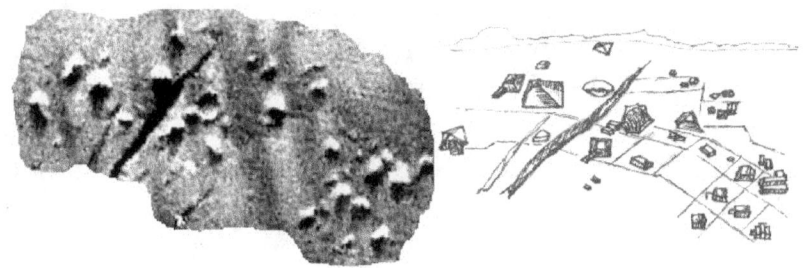

The next city seems to have a perfect stepped pyramid at the upper right. I have drawn its general shape to the right for clarity. Several steps and platforms are clearly visible.

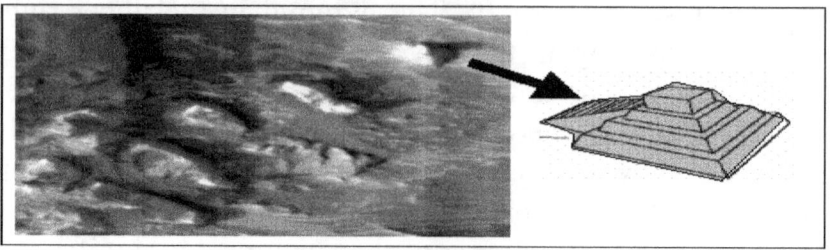

## Buildings Galore

Speaking of right angles look at this pentagon that was found on Mars. Several steps and platforms are clearly visible.

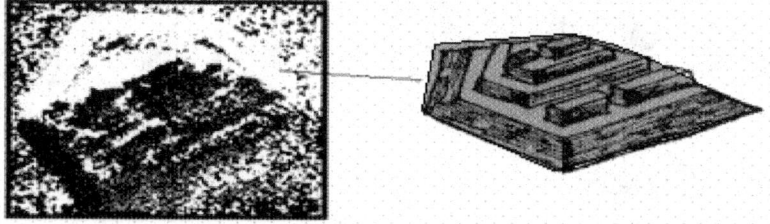

Can there be any doubt that these some of these structures are man-made? The Pentagon was first discovered by Richard Hoagland and it still impresses all who see it.

## Huge Surface Tunnel

If you say it didn't matter that there was water, its simply too cold to live there. The answer might again be just below the surface. What if they typically lived underground and had surface tunnels to allow comfortable transport. Let's look at the artifact [next left] which is part of the MOC image M1501228. Many believe this to be a huge tunnel that is about 300 meters high and 200 meters wide. It is ribbed for strength, but was breached a long time ago. Maybe the Martians also lived underground during the initial war years. Reflecting light off the surface of the tunnel is evident in the photo which suggests a very smooth surface. No one knows what the tunnel was used for, but not many dispute that this feature would be almost impossible to be produced from natural phenomenon. If this rupture had happened hundreds of thousands of years ago, the cavity would have filled in and there is clearly an open area below the covering, so the tunnel would have been operational during the time of the 1st and second creations of men.

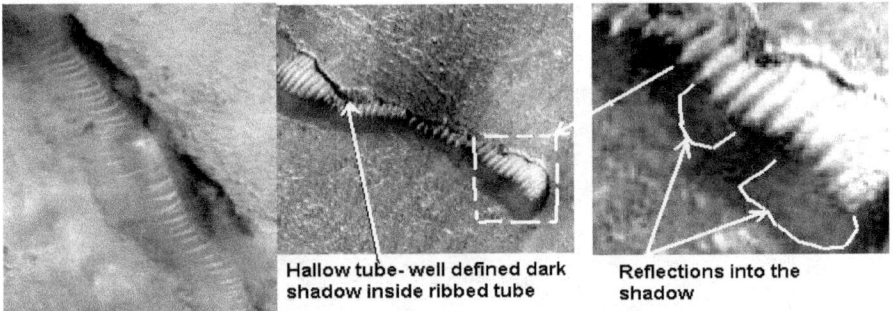

Hallow tube- well defined dark shadow inside ribbed tube

Reflections into the shadow

## Glass Martian Tunnel

Here is still another "tunnel" [above right] that has surfaced on Mars. Note the clear domed passageway as it can be seen in the fissure. If that's not enough, just about everywhere you can find these tonnels.

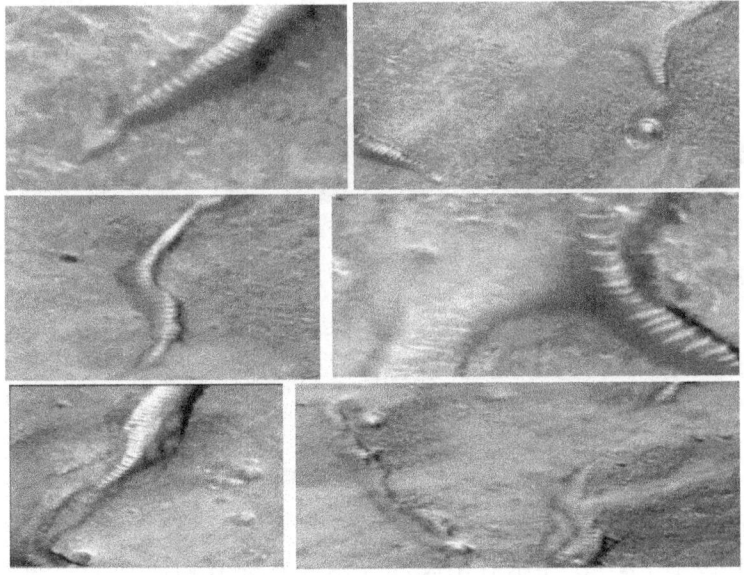

## The Famous Martian Airport

If those photographs are not enough to convince you; look the picture from Mariner 9 that Richard Hoagland termed as the Martian airport-out by itself, as you would expect, but with unbelievable detail. The, Tom Penner, rendering of the complex certainly shows that, at one time, this was an airport or huge shopping center or something else, but it was not an accidental structure. Actually the structure is mostly underground rather than above ground as the artist conception shows, but you can get the idea just the same.

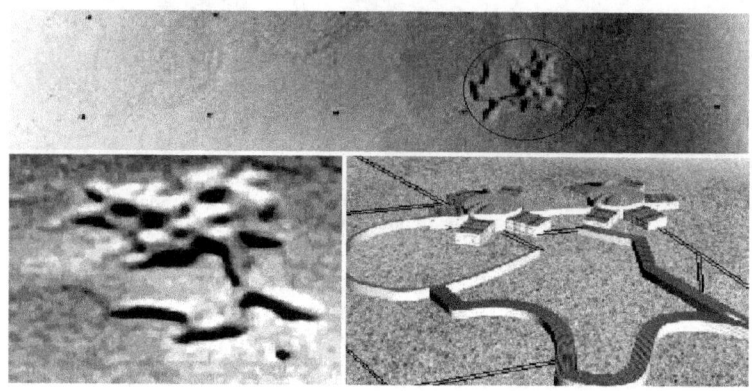

## Buildings with a Lot of Windows

In the following picture, the inserts for windows or whatever can be easily made out. These are not natural window cutouts of a rectangular cliff, but instead shows characteristics of being manmade.

## Martian North Pole Things

This is really something to see. When we look north, we find extremely unusual billboards or graves or markers of some sort. It is difficult to believe that these tall structures, shown below, could have been formed naturally. Whatever these huge billboard objects are, they are not covered over from thousands of years of existence and must have been made fairly recently. We could estimate that they have been there no longer than a million years ago and probably much less time.

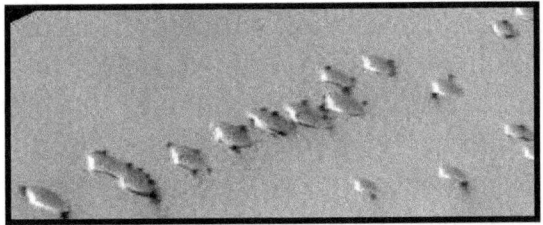

**Martian Roads?**

***Lines Mean Something*-**What are the lines in the Martian picture below? For miles and miles, traveling over the top of crater blast areas, the set of furrows on the Martian landscape picture extend in an absolutely straight line. No naturally occurring phenomenon can account for the lines that continue over mounds and through troughs and we know that they were made in recent times because the meteor blast marks are below the furrows and none are on top of the strange lines. It almost looks like a huge farm. The deepest furrows are towards the center of the picture, but you can see many more furrows below the major ones all extending in the same direction. No one knows what they are, but they were made very recently.

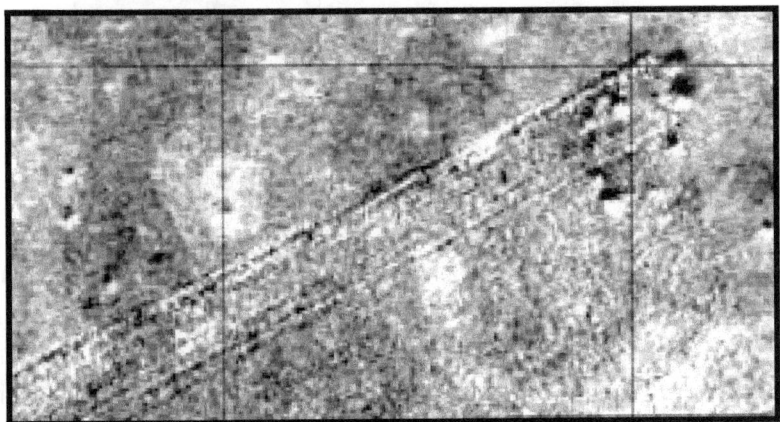

# Martian Water

Something happened on Mars that destroyed everything and soon the atmosphere left. That event was not the near collision with earth that left half of the planet blown away and the remainder pitted with craters. That earlier event, most likely occurred about 200 million years ago. Even after that catastrophe water was still abundant on Mars and with water there was, most likely a reasonable atmosphere. At a later date, the atmosphere began to thin and the water began to leave. Although we don't know the exact time period for this final disaster for any colonists, the picture below shows the remains of a huge sea on Mars. The remains are not very old and the details are still distinct, so we can be fairly certain that it was not very long ago. Some researchers place the rippling of the ground at about 17 thousand years ago by its strong definition.

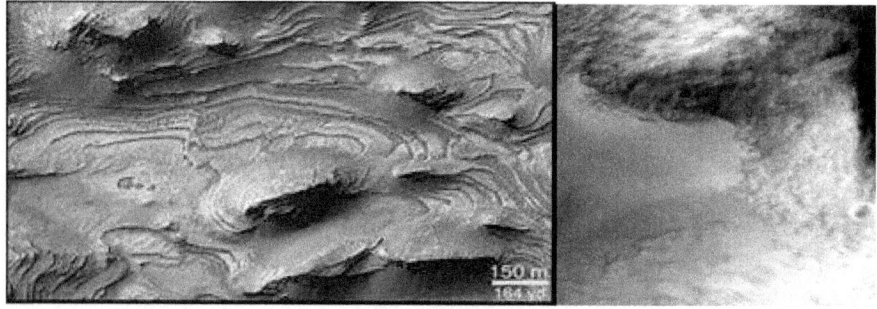

## Liquid Water

Speaking of the rippling of the ground and a 17 thousand year ago loss of water, here is another twist. From studies of Martian meteoric material, Dr. Leshin of the University of Arizona and others have concluded that Martian water originally contained higher deuterium levels than previously thought and the Martian atmosphere has lost two to three times less water through the eons than "dry planet" models suggest. Therefore some have concluded, there must be a huge ocean-like reservoir of water beneath the planet's surface. Not 17,000 years ago, I'm talking about water on Mars today. The recent picture shown above shows a reflection off of a shiny surface. It appears to be a frozen lake on Mars.

## Martian Sea Port

This area is commonly known as the port. Note the extremely regular shapes associated with the building on the cliff that once could have been adjacent to a waterway below. I have superimposed a drawing of the building for clarity. The right angled surfaces of the building are still very distinct showing that it was not abandoned extremely long ago.

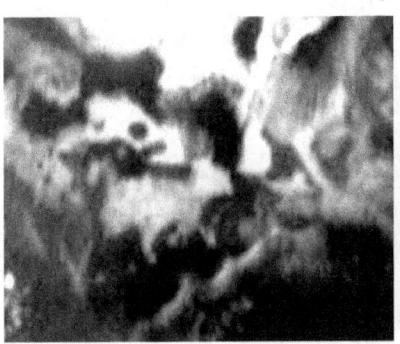 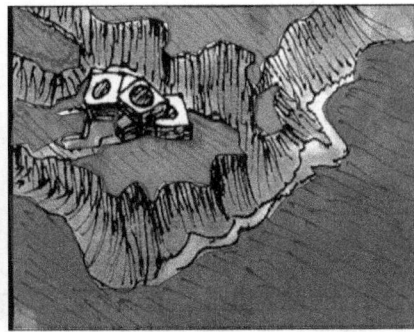

## Martian Terra-forming

The humans on Mars may have begun the repair of Mars in our fairly recent past or Mars may be fixing itself. It's been a long time since life could exist on Mars, but there is new evidence that life may be reemerging in the form of some type of plant life. Below is an area many believe is covered with some type of treelike plants. The color and density of these things change over a yearly cycle and they seem to have limbs like trees.

# Martian City Destruction

Not only do we find cities, terraforming possibilities, and manmade objects, we find evidence of war or something very similar. We see cities with complete destruction. Look at some more of the curious details found by the many researchers in this area. Can there be any reasonable doubt that these structures are regularly spaced, rectangular, and have high walls just like you would expect in a city. Even what appears to be sectored roadways can be made out. The most striking element is that the building appear to be melted.

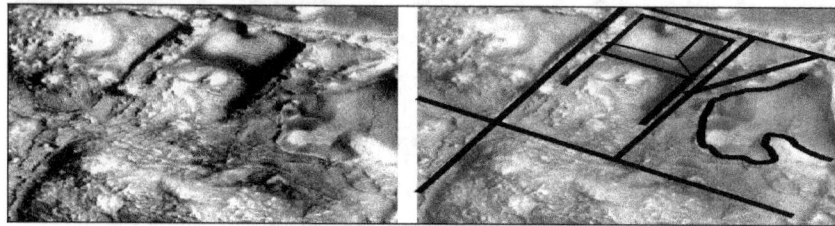

In the picture following, you can see the crumbled walls of the city pretty clearly. It looks almost like the city had been in a war.

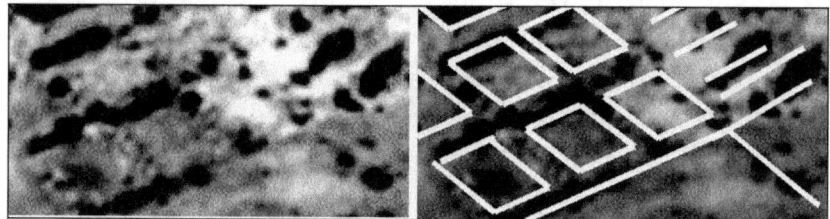

## Xenon-129 Evidence

This is the last thing I'm going to mention about Mars in this book. This Xenon-129 stuff is a "second order nuclear fission by-product" and guess where too much is found. Mars has nuclear by-product in abundance. The problem is that no one can determine how nuclear blasts went off on Mars. Don't believe stories of Mars collecting the material from some ancient supernova because it just doesn't make sense. What does make sense is nuclear bombs, but then there would have to be people there and someone must have been mad at those people. Many of the scientists trying to figure out why this Xenon 129 is there don't believe in the very recent nuclear explosions on Mars, but they are completely baffled as to what else might have caused the quantities of Xenon-129. That type of stubbornness keeps our history books comfortable and useless.

**Trees-**Speaking of trees, we can find some on Mars. The trees in this area seem to be barely clinking to life. Somehow the tops of these dunes allowed their survival, but we could hardly call this a forest. The next picture gets us closer. If we blow up a section, the trees can really be examined easily.

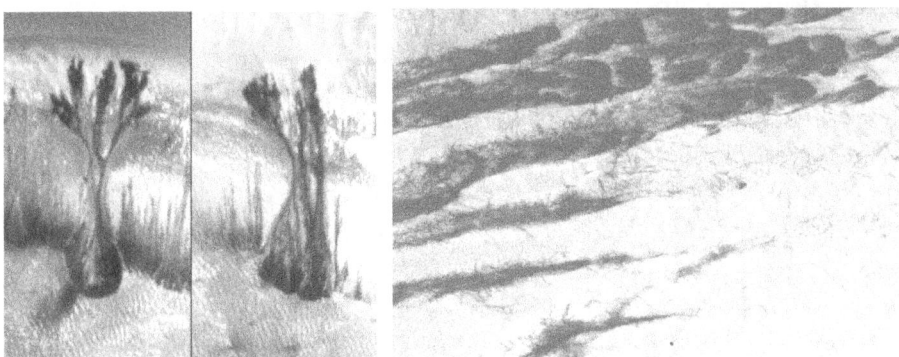

You know the old saying. If you have burned up trees, humans must have been there. While the trees look burned up, possibly that is the way tree look like on a planet with almost no air and almost no above ground CO2 respirators

**Flying-** We sent up a rover to investigate and what do you suppose it saw? The rover caught a Martin UFO, then another and another.

**Credibility Note**

131

Certainly some of the above examples are truly anomalies produced by wind, but the more of these things that pop up, the more it appears to be likely that the wind could not do them all. Some, must have been done by humans. I don't mean prehistoric, ape-like beings, I mean space age people.

## Timeline for Mars

The following is a possible timeline of the development and destruction of Mars.

***Years ago-    What apparently occurred***

**4.55 Billion-**The planet was formed

**440,000-** This was a first close encounter between Mars and Earth which produced the American Ridge Mountain ranges. On earth 90% of all life was destroyed. Mars could not have fared well.

**300,000-**There is a likelihood that Mars came close to Earth during this time period and established the Himalayan Ridge Mountain range.

**200,000-**Mars and Earth almost collide again and the Pacific Ocean was formed. Half of Mars split away. Particles became the Asteroids.

**150,000-**Cities were established on Mars by this time.

**120,000-**On Earth 85% of all life became extinct and huge meteors hits its surface. Mars would have had a horrible time as well. .The Atmosphere probably began leaving about this time.

**70,000 -**With the atmosphere all but gone, the water began to evaporate a go away from the planet.

**50,000 -**Some type of war destroyed the remaining outputs on Mars

Water that still remains on the planet, but it has been estimated to be only 10 percent of what it was in the not too distant past and with that water there would have been an atmosphere, plants, and almost assuredly it held humans.

# Life Beyond Mars

We won't spend much time on this subject, because not much is known about the history of the far away planets, but we should at least look at the possibility of life on other portions of our solar system. Certainly, the Jovian planets cannot sustain life, but what about the Moons of these monstrous balls of gas?.

Our solar system consists of sixteen large planetoids. Some of the "moons" are larger than Pluto. One of Jupiter's moons, named Granymede and one of Saturn's moon's, named Titan, hold claim to being larger even than the planet Mercury. These larger moons are big enough to have an atmosphere. If we can find water, it might show that they could have sustained life..

Some of the terrestrial planetoids are moons of Jupiter, Earth, Saturn, and Neptune. By far, Jupiter has the largest quantity of large planetoids around it with four, but let's start with Neptune and work our way towards Earth. Neptune's major moon is named Triton. Let's see if it has possibilities of sustaining life.

### Neptune's Terrestrial Moon

Any likelihood of Pluto or its moon having ever supported life is extremely farfetched and even Neptune is an extremely unlikely candidate. Neptune, as shown next left, is one of the massive ringed planets. Triton, however, is Neptune's moon of interest and we find a little bit of a surprise there. It's not formless as one might think nor is it just a chunk of rock. It has extensive ridges and valleys and appears to be inhabitable and if water means life, we really found a good place here.

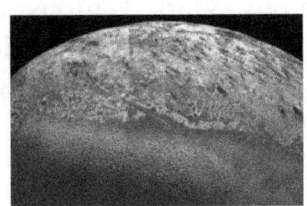

**We Found Water-**Triton is 25% water and is classed as terrestrial. It also has a mostly nitrogen atmosphere and has organic compounds existing on its surface. It has few craters and the surface has a changeable Polar Ice Cap shown in the picture. [Previous left] So here we have seasons, organic compounds and an atmosphere that is nitrogen based just like earth. All we need is heat that comes in the form of volcanoes found on its surface. One slight disadvantage is that the volcanoes expel water rather than molten material, so it's still REALLY cold, but the necessary components for life are there. How would you like to live on the planetoid above and look up at the planet Neptune? Let's come in a little closer to the sun and investigate further.

## Saturn's Terrestrial Moons

As we move closer to earth we go by Saturn. Somewhere in the mass of ice particles and debris that make up the planets rings are the 6 significant moons of Saturn. By far the largest one is named Titon, not to be confused with Neptune's Triton. It is still very far from the sun and therefore very cold. Titon has some possibilities when it comes to life as does a smaller moon named Enceladus.

## Titon

At first it seems like an ideal planetoid. It has some of the requirements of life and the really neat thing about this place is that it has a huge continent surrounded by liquid. Unfortunately, it lacks a lot, in my opinion and probably in the opinion of most organisms.

**We Found Water-**Titon contains water and is therefore considered terrestrial, however, it is not likely that it contained life. So you might ask, "Why?". It all has to do with the oceans on the planetoid. Dark regions on the surface indicate large continental masses are surrounded by massive oceans. The problem is that these oceans don't contain fish because the oceans are made of Ethan.

The Atmosphere is mostly Nitrogen like ours and organic compounds are in reasonable supply, so simple organisms could exist so long as they stayed away from the oceans and, of course, humans would not like it there without some major thermal gear. As pictured below left, the terrestrial planetoid known as Titon is, most likely, empty, but complex chemistry and liquids on its surface place it higher on the list of possible life supporting planetoids. The close up photograph of Titon makes it appear to be very inviting as the oceans gently ride next to the Titon beaches.

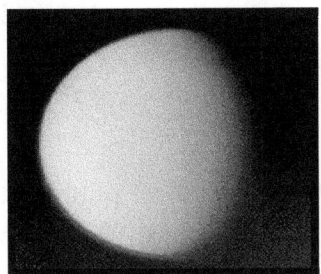 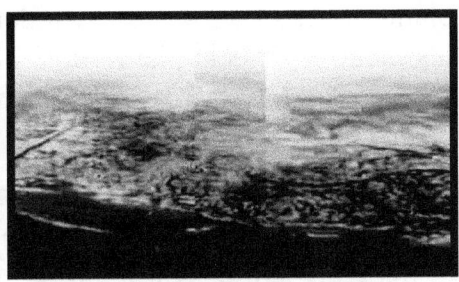

Besides Titon there is other hope of finding where human space travel might have deposited an outpost of travelers in the rings of Saturn as we look at the planetoid known as Enceladus.

## Enceladus

On this strange planetoid there are linear sets of grooves tens of kilometers long that traverse the surface. The uncratered regions are geologically young and may indicate that Enceladus has experienced a period of relatively recent internal melting. While it is the smallest of the 6 major moons of Saturn, it is, apparently heated by reactions with its closely linked cousin Dione. The satellite is about 500 km in diameter and is shiny. When I say shiny I mean it. It is 90% the brightness of a mirrored surface, so it is surfaced with a beautiful layer of snow.

If we look at a close up of the planetoid [See next right], we find that around some of the volcanic craters are refrozen areas that were once liquid water. It may still have liquid water near its surface which makes it a good candidate for supporting life. When a planetoid has enough heat to do that, it can sustain life.

No other signs have been found, but it is way out there and it's hard to get reasonable detail.

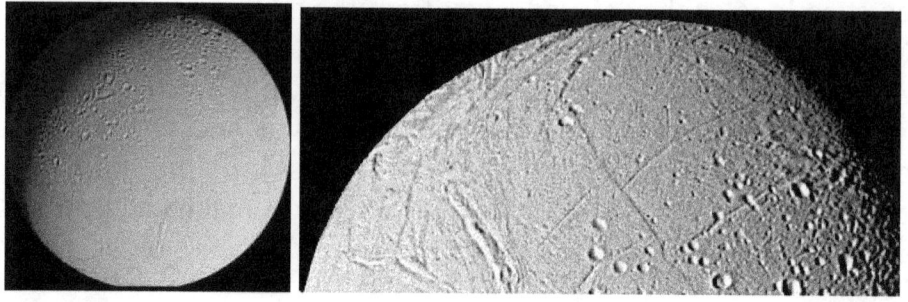

This planetoid is like a winter wonderland as shown above. Even with its heating mechanism, this like all the other moons of Saturn, is extremely cold, so let's go in towards Jupiter to look for signs of human space travel.

**Others**-Before we do, there are a couple of other planetoids to investigate. The first is called Iapetus and it is shown on the left in the next grouping. It's strange in that half of the surface is dark while half the planetoid is white. No one knows why it is this way, and I don't know if it has anything to do with space travel. It's just weird. Speaking of weird, how about an eyeball in space. If we look at Mimas, the thing that comes to my mind is a huge eye. I know it's a huge crater from an impact that almost split the planetoid apart, but the eye thing keeps getting in my head. Both of these places are extremely cold and presumably lifeless.

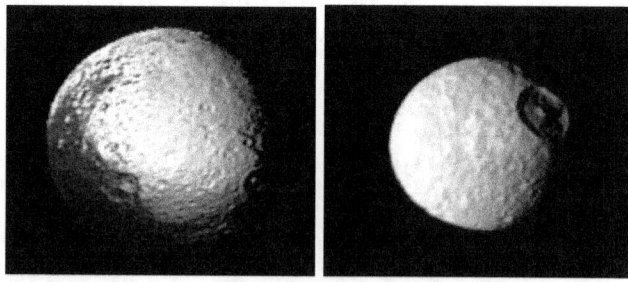

### Jupiter's Terrestrial Moons

**We Found Water Again**-Jupiter itself is no place to have humans, but some believe that this monster has a reasonable high probability of supporting some type of life. I don't see it, but its 4 major moons are huge planetoids which contain the main requirements for life. They contain water and they contain Oxygen. Europa, Callisto, and Granymede are classed as terrestrial because of these features, but they all have one serious flaw. Like the earlier planetoids mentioned, they are COLD. One of the moons is fairly hot so you would think that would be a winner, but it has almost no water. That moon is Io. Don't count Io out because heat is pretty valuable out in the far reaches of our solar system. With four of Jupiter's giant moons having what is necessary for life, the question still is, "Was there ever life there?" Even though you will never get most Astronomers to agree, the answer could be, "Yes!" Here are some pictures of these places that could show elements of life. If life had been on these planetoids at one time, they certainly would have been hardy, because even the hot lava flowing from their centers freezes very quickly after reaching the surface.

**Io-**Io looks weird. It has a sulfur dioxide atmosphere and not much water. Some don't consider it to be terrestrial, because the surface is covered with silicate lava flows from the huge erupting volcanoes. I brought it up because it is huge and it has something special. Io is warm enough for life and its complex chemistry make it suited for establishing some forms of life. Look at the eruptions that extend hundreds of miles into its atmosphere. [See below right] Humans could have stayed here for short periods of time.

**Callisto-** Although it has water, it is all frozen. You can see from the cracked surface that it is almost completely covered in ice. There seems to be no volcanic activity, the atmosphere is mostly carbon dioxide, and there are craters everywhere from meteor strikes. [See following middle] It doesn't appear to be inhabitable, but there is something curious that was found.

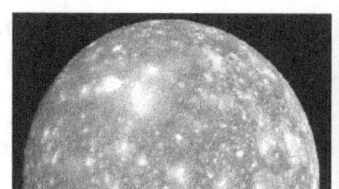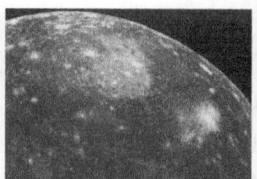

Here's the strange thing. Just like on the planet Venus, we have found a long line of blast craters. [See above right] It is unlikely that meteors could have pelted the surface in line as depicted by some natural phenomenon. Why would someone blast an empty planetoid? Some try to indicate that meteors hit and then bounce, but in this case the meteors would have bounced about 30 times to make the feature shown---Not Likely—and Yes! The string of craters does get larger in the middle and smaller on both entry and exit. [Not that is an unusual bouncing object..]

**Granymede-**Granymede resembles Saturn's moon named Enceladus. Like Enceladus, it too is terrestrial and could have had an outpost stationed there for some time. Like the indications on Callisto, there are even indications that there were battles fought on this distant planet. Granymede may have been farmland of some kind in the distant past, but it doesn't have any recent volcanic activity that could increase the surface temperature, so it's really, really, really cold. Let's look at pictures that may indicate ancient life may have once existed that didn't mind the cold. Although Granymede is cold and has minor amounts of oxygen just like Callisto and Europa, its topography is unusually structured. Huge blast craters have dust fingers out to over 600 miles and light and dark areas seem to show areas of different temperatures. [Below middle]

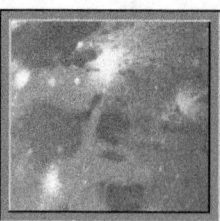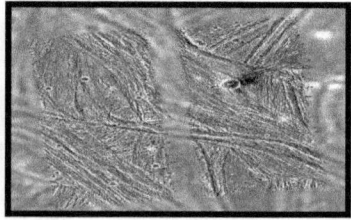

The most remarkable thing on Granymede is the long parallel lines that can be found at various areas. As we look closer at the "lines" we start to see an unusual structure. [Above right]

Look at the following picture [Left] showing the regular furrows and the changes in direction all up to the mountains pictured on the left. I have no idea what could be farmed there, but the area looks very much like farm land. I don't know what natural event would have made the right angles and line straight furrows. Do You?

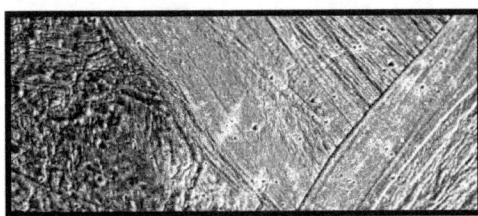 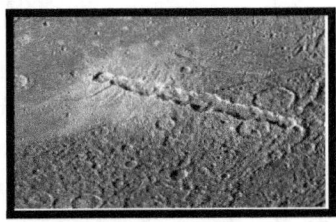

Here's another strange thing. [See Above right] Just like on the planet Venus, Callisto, and even our own moon, we have found a long line of blast craters on Granymede. The strike path doesn't come from a meteor that split up just above the ground in even pieces, nor did a meteor go bouncing around 16 times. What do you think these evenly spaced almost identically sized blast areas are? It should also be noted that this feature is not very old as all of the other features in the area are under the 16 blasts. It also should be noticed that the center crater is larger than the entry and exit ones like we found on Callisto. Another thing that is similar to the artifact found on Callisto is the fact that the center blast is stronger than those before or after the mid-point. Does that sound like the breakup of an asteroid???????

**Europa**-*Europa* is the smallest of the 4 major planetoids of Jupiter and perhaps the most interesting. It is classed as a terrestrial planetoid as well and, this time, with good reason. This place may very well have sustained life at one time and there may be reasonable evidence in the very detailed pictures we have during our flyby excursions. One thing that helps Europa is that it is closely united with Io. The forces of attraction between these two planetoids may cause heat and you really would like some heat when you are stationed on a planetoid around Jupiter, cause it is cold out there. Europa is a really strange place. Look at the fingers [next middle], These travel around the entire surface, but no one really knows what the streaks mean nor does it mean there was life.

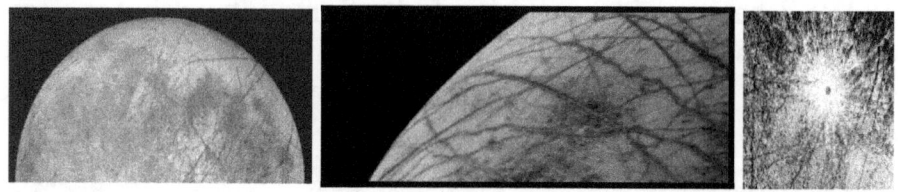

Volcanoes tell us about the inside of the planetoid. Unlike its sister that has molten Ice coming erupting, this one has lava. [See above right] That, of course means that the inside is substantially warmer than the outside which would also aid terrestrial life. From the lava signature we now know that the center is molten iron and sulfur. Volcanoes could have been a heat generator that aided human survival in the past. While the volcanoes aided life, meteoric blast areas, as shown, spread material out hundreds of miles from meteorite strikes. These massive strikes would have made it very difficult to stay for any length of time.

Tests on the atmosphere show it to be the only other planetoid to have <u>real uncombined oxygen in the atmosphere</u> besides Mars, Venus, and Earth, so let's add breathable air.

Researchers now are almost certain <u>that liquid water</u> is causing the surface ice ripples to make the circular patterns found around the planetoid. The picture below shows only one of the many areas that have these curved ice cracks. Under the ice is warm--- ok! Ice cold—but liquid water.

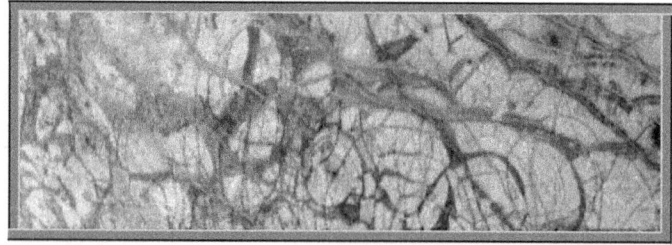

If you have oxygen in the atmosphere, water, and enough geothermal heat to liquefy the water. You can have life. What is the first thing live things do?---Of course you know!--- they make roads and cities.

**Europa Roads**-The most amazing things to see on Europa are its roadways. These roads, somewhat similar to the long trails found on Phobos, are everywhere. They travel hundreds and hundreds of miles without varying their width. On either side of these roadways are high embankments so they could not have been waterways or anything else that could have been natural, but they are there just the same. OK! Some try to say that the dual parallel lines are continuous cracks produced by the uneven heating of the planet, but the problem is that these uneven crack go on for miles and miles as though massive block of ice with exactly the same temperature broke away from other massive ice blocks that were also evenly heated and had similar thickness over the entire separated block and these "cracks" are not jagged, but, instead, conform to the natural lay of the land.. I'm sorry to say, that doesn't make any sense to me.

In with all the roadways, are sections of geometrically shaped and squared structures which have been thought to be the remains of cities covered over with ice. Naturally you need cities to go with roads. Notice how the "roads go around the hills. I've highlighted what I'm talking about here.

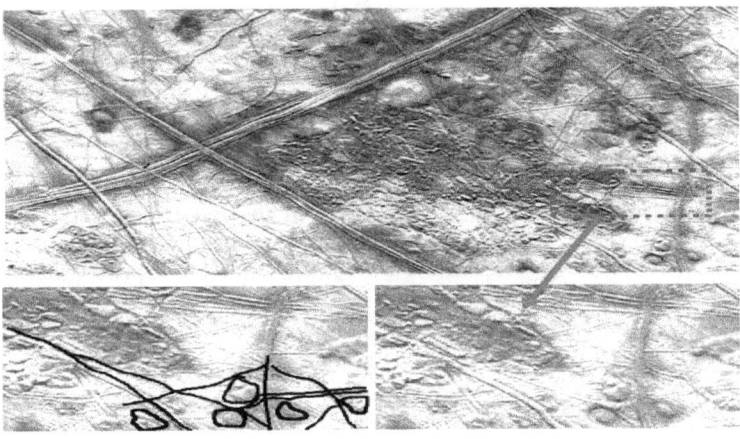

There is no certainty that any of the Jupiterian planetoids ever had life, but, with the building blocks of both water and oxygen, they do raise an interesting question concerning other humans in our solar system living beyond Mars. Europa seems to be the most likely place if there is one. People living on these planetoids today is unlikely, but that does not mean that there were never inhabitants nor does it mean that no creatures exist besides those on our planet.

## The Moon

We made it through all of the probable planets that could have had outpost of humans in our distant past, but I skipped over the Venus because its environment changed very recently and I skipped over the moon for another reason. It has the highest probability of having humans on it simply because of its location. The moon has its own peculiarities and some theories should be, at least, presented to account for the anomalies in both text and physical evidence. The moon doesn't have air, but, there is evidence to show that there was an outpost on the moon during the ancient times. Ancient texts talk about battles in flying ships on the moon, but only recently we have found what are the remains of the areas where the battles may have been fought.

**Huge Tower-**Below are pictures taken during various flybys of the lunar surface. Look at the immense shaft rising from the surface that was seen during the Lunar Orbiter II flyby in 1967. The frame number is LO-III-84M for those who want to check it out from other sources. Some tried to explain it away as a shaft of gas erupting from the surface, others said it was an optical illusion. Look at the shadow of the illusion and note that gases would have been dispersed as it rose from the surface. This shaft is solid almost 2.5 kilometers high. The shaft is probably man-made and taller than a 700 story building. I don't know what it is, but it's not an ordinary rock.

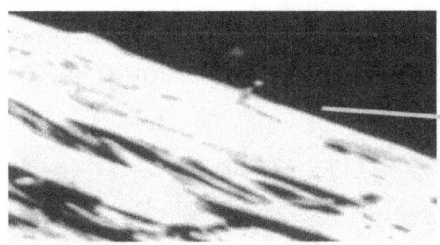 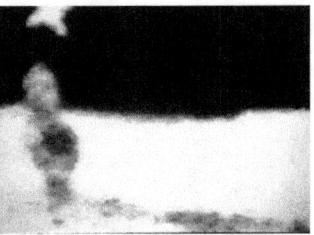

**Microbes on the Moon-**Here is some interesting data. Scientists have now concluded that the jumble of circles below prove life existed on the moon at one time. The structure is identical to fossilized microbes found here on Earth. Microbes don't just appear, something living has been on the Moon at one time.

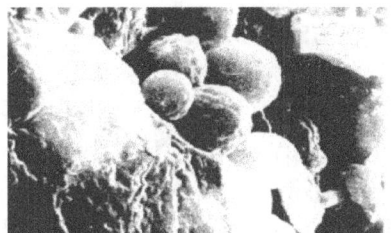 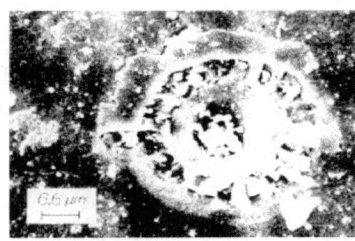

**Blast Holes-**There is another group of features that are very strange on the moon. Those are long straight sequences of blast holes. Some have tried to explain them away by saying a rock hit the surface and kept on bouncing for tens of miles. Although the rock hop thing sounds ridiculous, another alternative is a bombing run and no one wants to say that so the hoping rock stays on top. Just imaging a huge meteoric rock bouncing as many as 26 times before finally landing and then it gets even more bizarre as the high bouncing rock simply disappears. Below are two areas on the moon with this type of blast evidence, but there are many more.

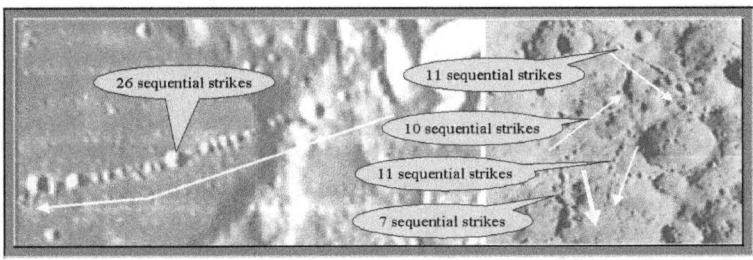

**Close Proximity-**One good reason for the inhabitation of the moon is that it is close to Earth and supplies could be quickly retrieved.

As the Earth rises on the moon, its closeness would have comforted the humans stationed there. We will revisit the moon with still more evidence of life, however, what we will find later is that life may still be there and still responsible for flying. Quit laughing until you read the later section. The Moon may have played an important role in two major wars. One occurred about 11 thousand years ago and a second occurred a mere 6 thousand years ago. While this book is not getting into the details of these horrible times on earth, it should be noted that there is considerable evidence of the wars and reasonable evidence of the participation of the colonists on the Moon, Venus and even possibly even Mars. The moon possibly was an interim stop for attacking and defensive type vessels just like you have seen on science fiction movies. The reason we should believe this is that ancient texts, some of which I have already presented, tell about its involvement directly and many strange artifacts have been found on the surface, which could not be natural. Some more evidentiary objects are shown next.

**Similar Landscape-**While the majority of the Moon looks bleak some parts of the moon look very similar to here on earth. Rolling hills and a meandering "dry" riverbed make the picture look like the moon is a perfect place. We can even make out what appears to be roads that take right angle turns. We know there is a shortage of air and water, but both could be manufactured from elements on the surface and people could be fairly comfortable living on the moon as sort of a faraway space station, especially with the rolling hills shown below.

**Possible Shopping Malls-**Then we come to flowers, or what appear to be flowers. These strange objects are all over and don't appear to be natural, but I don't know what they are. Four of them are shown below. I have tried to draw them in more of a 3D fashion below the photographs, but it is not known what they really are. They have the general appearance of modern shopping malls or airport.

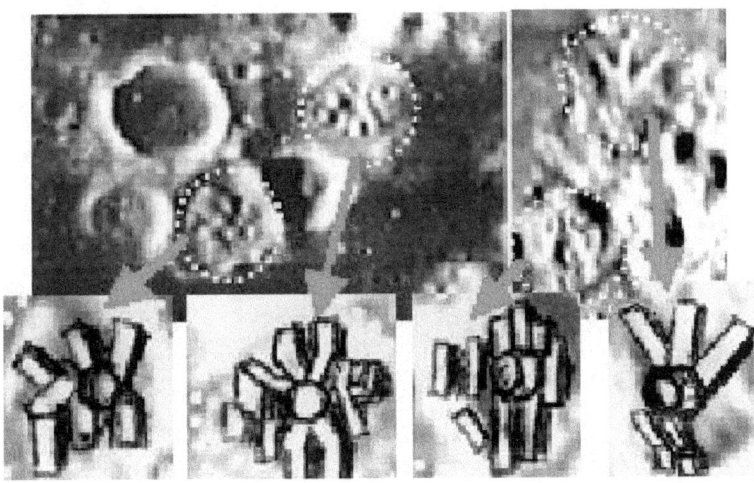

**Walking Evidence-**If the humans once lived on the moon we might expect to find holes. If we look at the Clementine lunar photograph, LO-III-123-H1, we find something very strange. It is shown next. The surface around this area is covered with strange double-hole patterns. No one knows what they are, and they have not been satisfactorily identified as natural. Maybe these things have something to do with those trying to live on the moon.

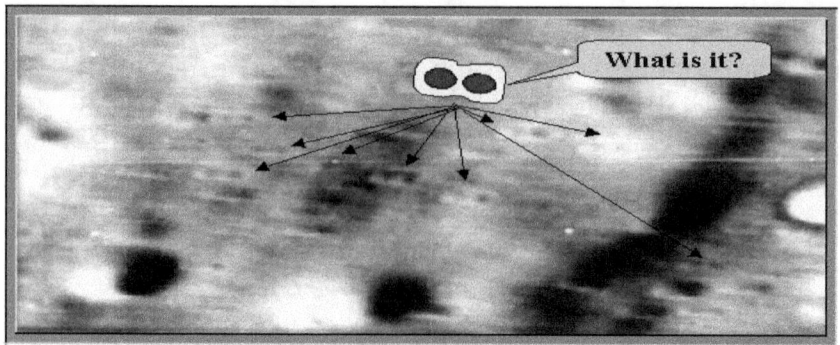

The patterns show that whatever was "walking" went in a walking pattern and shifted its weight from back to front just like we do when we walk, however, these "footprints" are much too large to be from humans. They would have been made from some type of walking machine. You can see that they are all over the place, so whatever was walking did it a lot.

# When Were the Planets Inhabited?

From the evidence, we can assume that people lived on some of them at one time. Now another question comes up and that is, "When did the various planetoids have human outposts on them?" For the answer, we would need to look at the wars mentioned above.

**War Evidence-**While not presented in detail this book, there are many ancient documents confirming huge wars that occurred about 12 thousand years ago. I don't mean the rock throwing kind either, Some of the texts talk about flying machines, deadly blasts that level entire cities and other "hard to believe" elements. There are also discussions of fighting on the moon and worldwide warring. Just because they are hard to believe, many ignore the texts as fanciful, but the sheer quantity and detail make it hard to believe these were fantasies. The evidence is not locked up in stories, but also physical evidence of wars has been found throughout the world. The evidence includes underground cities around the world, blast evidence with intense heat to turn sand into glass, fort walls demolished and the same "glass" effects, radioactive bodies found in piles. The group of radioactive remains next was found in ancient Mohen jo Daro in Pakistan. The bodies are still radioactive today and guess when they became radioactive. Give up?

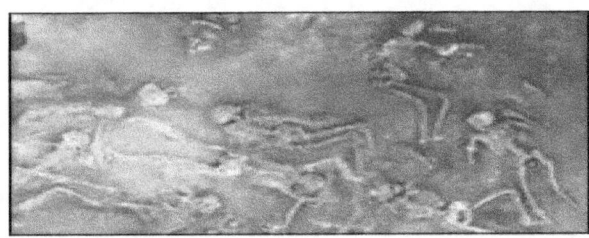

The answer is that all we know is that it happened over 5 thousand years ago. We might be able to find clues to connect the earthbound verbal and physical timing with disasters on the planets, if fighting on the earth in some way affected the nearby planets and their colonists. I know this all sounds like I'm talking about little green men on planets visiting earth, but I'm not in fact, there is little evidence to suggest that humanoids or other creatures live on these planets today. The collected evidence only suggests civilizations were there thousands of years ago. The apparent buildings and artifacts found on these planets could have been left over from just about any time as no major erosion event was on any of the planets besides Venus. Two things give us indications of age: water and volcanic action. If there were no water now or in the relatively recent past, that would be a good indication that the artifacts were of an extremely distant past. If the planet surface had extreme volcanic disruption, the artifacts would be destroyed very quickly. Here come the scientists to tell us more.

### Martian Dirt

The Martian soil may give a pretty good clue as to when the destruction of the atmosphere and the major portion of the civilization by examining the water supply. It has been determined that some of the dry riverbeds were wet as recently as **20 thousand years ago** on Mars. Below is one of the probable riverbeds so the answer on that planet is any time before then.

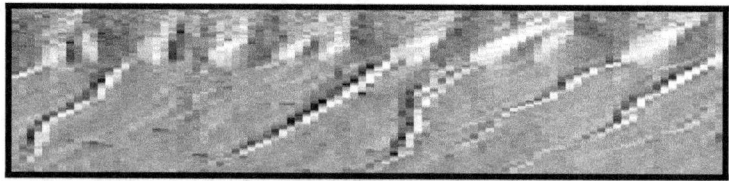

### Venusian Soil

Venus gives us a similar determination and time period for the loss of water. River deltas are found around the planet, but, in this case, all the water died up very quickly in the recent past. Below is one of the dried up river deltas found. This quick dry time **could not have been hundreds of thousands of years ago** or the details of the riverlets would have been consumed by the planetary actions, erosion, liquification and other tectonic occurrences.

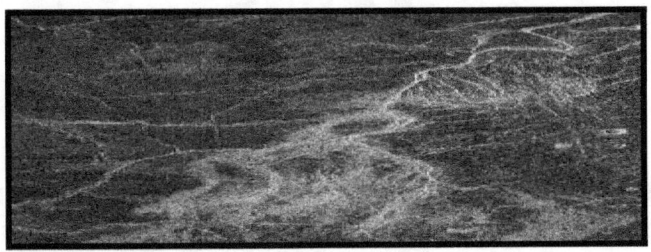

## Lunar Water

The Moon could never have been inhabited unless there was water. Sure enough, scientists now know that water is still located at the North Pole. No reasonable dating can be made, but life could have been possible with water. Because the water is still there, we can believe that habitation could have been **in the not too distant past.**

## Mercurian Water

On Mercury they found water on both poles. More water is located on Mercury than there is on the Moon. Because its rotation and revolution cycles are so similar that there could have once been life. At the poles there could have been some semblance of life provided that the population slowly migrated to insure that they stayed on the "dark" side of the planet and along the pole. Just think of it. There may have been life on all of the inner terrestrial planets. This life may have been present **as recently as 10 thousand years ago.**

## Europan Water

Scientists have also found an abundance of water on one of Jupiter's moons, Europa, which makes it also a strong candidate for ancient life. Other planets also have various amounts of water. That brings us back to Venus who fared worse than any planet during this fairly recent epoch in planetary history. The indication is that Venusian life was great up until **11 thousand years ago.**

# Venus

That brings us to the last of the planets of interest when it comes to habitation. It is likely that Venus was a very inhabitable place in the very ancient times, but some bad things obviously happened to it in the very recent past. As I previously showed, the orbit of Venus was much closer to the earth in the ancient times. Many writings discussed the huge planet in the sky. After the close encounter of earth and Mars about 400 thousand years ago, earth's orbit was pushed closer to the sun and closer to that of Venus.

**Venus and Earth Close-**For many years the two planets were so close, that Venus looked like a huge orb in the sky. Many of the ancient works tell us about how our ancestors saw or had heard about the original placement of Venus in the sky. With this close orbit, it would have been very reasonable for the people of that time to visit Venus.

**Ancient Flight-**While some may not know about the ancient capabilities concerning flight, let me say that there has been so much written and so much physical evidence to prove that flight was commonplace during the very ancient times that an many people today have written complete books on this subject alone. I also wrote a book on the subject. Humans flew and there is reasonable evidence that they flew to Venus and other planets.

**All of a Sudden-**From out of nowhere, it seems, something terrible happened. The Venusian moon exploded. We know that something exploded very close to the Venusian surface and we know that earth was very close to the planet when the explosion happened. The important thing to know about the explosion is that it happened only 12 thousand years ago. That event and the details around it are called the Venusian Moon Theory.

# Venusian Moon Theory

This theory will seem bizarre without some reasonable backing which will follow. Before I get into the details of how we can be assured that this happened, here is the theory in a nutshell. The Earth was somehow instrumental in causing a fairly large moon of Venus to shatter.

**Here are the Elements-**Because the Earth somehow initiated the Venusian moon shattering, the plasma streamers between the two planets were established that are still in existence today.

- *The ancient inhabitants of the Earth saw and recorded what appeared to be a huge comet because of the Plasma trail between Venus and Earth. Venus was quickly given the name Wavy haired or flowing haired planet because people saw the hairy plasma trails.*

- ***The Venusian moon's rotational axis*** *was around the equatorial region of Venus, so that was the only region assaulted by the meteoritic particles from the blast. The moon was close to the planet, so the meteors could not move away from the revolution path of the now demolished moon before they hit the Venusian surface. Obvious evidence was left behind.*

- ***As Venus spun around***, *the particles kept on coming forming the ring of craters and that horrible gash seen today as Venus was almost split in half during the assault.*

- ***This onslaught initiated substantial volcanic*** *and other atmospheric disturbing actions, which put the planet into its current over heated, heavy atmosphere state.*

- *Before the Earth rotational axis changed* to where it is today, the particles from the explosion began reaching the Earth by the thousands and began hitting areas around the Earth generally in an area confined near the equator. *[Again, the evidence is easy to find..]*

- *The sky would have blackened* before, during, and after the assault as indicated in ancient texts, because of the dust from the meteorite cloud.

- *The darkened sky would have thrust the Earth* into a time I call the Venusian Dark Ages and the nations of the world would have been weakened and dark. Things would not have grown and people would have starved by the thousands.

- *The orbit of Venus* could have been repositioned slightly closer to the Sun as part of the event, which would have pulled the Earth to a slightly closer orbit as well.

- *With the earth slightly closer to the sun*, less of the water would be frozen and the Ocean height would begin to rise to a new average level. Venus, on the other hand, would become an inferno.

- *Island nations, like* the possibly not so mythical Atlantis, would have slowly submerged into the oceans as the water levels rose over the next few years.

The above theoretical event listing is the simple version, but I'm not trying to prove things here. What I'm trying to do is to open your mind to possibility so that by collecting enough cross compared bits and pieces. **Please Prove me Wrong!**

**Venusian Catastrophe**-Let me start off this section by giving a brief overview of some of the "passed down" information from eye witnesses. I'm talking about what the people living 12 thousand years ago saw. We might assume that the event was not well known during ancient times. We can only find written information about this event from Europe, South America, North America, Central America, Egypt, Sumeria and the rest of the Middle East, India, China, and the Pacific Islands so we shouldn't believe it happened. Some might tell you it was worldwide mass hysteria, but I have a hard time ignoring just about everyone writing about this time period. On the off chance that you might believe, here are a few of the written testimonies. What you will find is that there is a common theme.

- *The planet we call Venus* used to look a lot larger in the earth's sky.
- *The planet looked comet-like* during this time period. It had a tail that could be seen easily by all observers.
- *The planet had some major catastrophe* on it which made it appear to explode
- *Pieces of debris from the explosion* pelted much of the world.
- *Many places caught on fire* and many people died as a result of this assault on earth.
- *Venus appeared to catch on fire* as well and changed its position in the sky.

## Written Evidence

The fact that there was a terrible Meteor storm witnessed on earth was captured in a few ancient histories. Here are some from Europe, South America, North America, all over the Middle East, the Far East and the Pacific Islands.

**European Venusian Meteors-** The European countries remembered the devastating event and wrote about it. Greek stories are filled with details. Here is one.

*According to Greek legend, "A blazing star almost destroyed the world with fire before it became* **Venus.*"* **[Although it is difficult to interpret, here is what I believe they are trying to say. I believe they are saying Venus used to look like it had a blazing tail, something happened to it which almost destroyed the earth and finally, it settled down to become the Venus we now see.]**

**South American Venusian Meteors-**The People of South America remembered and wrote about it. The Inca legends tell the story.

*The Inca called **Venus** the "Wavy haired planet"; [**This also seems difficult to interpret. Could wavy hair be flames shooting from its surface during a time when the Inca were around?**]*

**Central American Venusian Meteors**-The People of Central America remembered and wrote about it. This is from one of the Aztec legends.

*The Aztecs called **Venus** "the Star that smoked" and said that it once passed by the world blazing and killing many people. The Aztec god, Quetzalcoatl, associated with Venus, is typically pictured with a wavy headdress. [**I'm going to get into this whole wavy haired tail thing in a minute, but the blazing and killing sounds like meteors hitting and setting fires. The meteors came from the smoking planet.**]*

In the Mayan Dresden Codex, the god of Venus is depicted with shooting darts. It seems to me that if something is shot away from a planet, it would have been meteor-like. The picture below is from one of the Dresden pages. You guessed it Venus is in the middle.

**Blackfoot Indian Venusian Meteors**-The People of North America remembered and wrote about it. Let's see what the Blackfoot had to say.

*According to their traditions, "The morning star [Venus] put on a scarlet cloak [**sounds like it was on fire.**] and appeared before a woman on Earth that he loved. She went into the sky with him, but was warned never to look back. She did, of course, and was ordered to return to Earth." [**The return was a mess if we believe the other histories.**]*

**Ute Indian Venusian Meteors**-The Ute Indians tell us the same thing in their verbal history.

*The sun was slivered into a thousand fragments, which fell to Earth causing a general fire. Then Ta-wats fled before the destruction he had wrought. All were consumed; until at last, swollen with heat, the eyes of the god burst and tears gushed forth in a flood which spread over the Earth and extinguished the fire." [**This flood is not the worldwide flood we have all heard about, but it was significant, just the same. As far as the sun bursting, I personally believe it was Venus and not the sun.**]*

**Egyptian Venusian Meteors-**In Egypt, the event was known and written about.

*Sonchie, the high priest, told Solon, a Greek historian, about events before the flood. He wrote, "Many are the destructions of mankind that have been and shall be. The greatest are by fire and water. During long intervals there **are deviations of the bodies that move around the Earth in the heavens** and the consequence is widespread destruction by fire of things on the Earth." [**The fires must have been everywhere when the Venusian moon split apart. The comment that it was one of the "Normal bodies that moved around in the Earth sky" limits the body to one of the close planets. Of course, the closest is Venus.**]*

**Jewish Venusian Meteors-**The Jews wrote about the event in the book of Enoch chapter 85 verse 1-4 we read:

*"A single star fell from heaven- raised up and fed among the cows--I saw many stars which descended and projected themselves from heaven to where the first star was." [**Some claim this and similar verses are figurative and depict Satan being thrown from heaven, but sometimes people simply write what they want people to read.**]*

**Sumerian Venusian Meteors-**The Sumerians made record of the blazing tail of Venus. Their goddess named Innana was associated with Venus and the information is the same as recorded by all the rest.

*"To the queen of the heavens Inanna [Venus], to her who filled the sky with her pure blaze. The luminations are as bright as the sun. Who initiated the flood-storm? You roared in the heavens*

*and Earth. You smote the flesh of the people."* **[The blaze of Venus filled the sky, roared across Earth and smote the people. I think the only way Venus could smote the people is if its moon exploded and pieces fell to earth as a huge meteorite storm.]**

*"She [Innana/Venus] who causes the heavens to rumble. She who shakes the Earthquake. She cried toward heaven and Earth, "My hair will whirl in heaven for you." You flash like lightning over the highlands. You throw firebrands across the Earth. You split apart the mountains.* **[The hair extending sounds like a reference to a comet-like tail or a blasted away section of Venus that hit the Earth. Firebrands hitting the earth sounds like meteors to me. Your firebrands might be different.]**

**Assyrian Venusian Meteors**-Assyrian literature tells the same story. This time the Venus-goddess is named Ishtar, but it is the same.

*"To the pure flame that fills the heaven, who shines like the sun 'Ishtar" [Venus]—"I ran battle down like flames in the fighting. I make heaven and Earth shake. I trample the Earth. I destroy what remains of the inhabited world".* **[To destroy the remains of the inhabited world, there must have been something substantial that happened with Venus.]**

**Arabian Venusian Meteors**-Coptic texts date the event for us in the Age of Leo. The ancient Arabic text called "Bundahishn" tells us the following:

*The Ancient Coptic text tells about a great fire and flood coming out of the constellation of Leo.* **[This not only describes the event but places it in the "Age of Leo", 11 to 13 thousand years ago.]** *It goes farther indicating that the beginning of world history was around 12 thousand years ago and some of the major deities were born during this event.* **[The beginning of history must have meant that there was a destruction period just before that time.]**

**Phoenician Venusian Meteors**-Phoenician texts describe the event, but this time the Venus-goddess is Astarte, the Phoenician version of Ishtar.

*"See, Astarte" [Venus], she descends into a pool as a fiery falling star".* **[A beautiful description for a meteoric terrible disaster.]**

**Persian Venusian Meteors**-Mandaean Texts from Persia give us the same information.

*"150 thousand years after man was created, the whole Earth broke out into flames and only 2 escaped."* It continues by saying that they had children and, of those ancestors, Noh [almost like Noah] was the one that survived the Flood that followed. **[The Earth being filled with flames could have been from the huge quantity of meteors from the explosion, but clearly this event occurred well before Noh survived a worldwide flood and sure enough, the worldwide flood evidence shows that it occurred about 9 thousand years ago, several thousand years after the suspected problems that Venus was having.]**

**Indian Venusian Meteors**-The Indian writers also informed us of this terrible calamity. The people remembered Venus sweeping away the stars.

*Indian literature states the following, "Her [Venus's] anger grew so terrible that she transformed herself, grew smaller and black. On a blind rampage she was killing everything and everyone in sight. **Her hair is wild**, her eyes red. The world trembles and cracks under her tread. Her dark hair flies in the sky sweeping away the sun and stars."* **[Again we read about the comet-like tail and so many meteors that the sky is darkened.]**

**Chinese Venusian Meteors**-The Far East writers also informed us of this terrible calamity. The people remembered Venus sending down a huge meteor shower.

*The Chinese writers said the same thing, "There was a time when a planet [Venus] approached close to the Earth, causing great showers of stones."* **[Not too many of the planets could have come close to earth. The moon of Venus is my guess.]**

*Venus was depicted as a dangerous "fire spitting planet" according to other Chinese legends.*

**Pacific Island Venusian Meteors**-Even the people of the Pacific remembered Venus sending down a huge meteor shower.

*Venus was depicted as a dangerous, "fire spitting, planet" by the Samoans. [**It is like reading the Chinese version. What would have given them that idea?**]*

## Physical Meteoric Evidence

Not only did people write about the event, but also, there is a huge amount of physical evidence in the form of thousands of craters left when the flaming masses hit the Earth. Pieces of material from the explosion hit places around the world. We know when they hit, we know that the explosion that caused them was reasonably close, and we know that there were many thousands of meteoric chunks that hit the earth at the same time. Here are some of the things we know

- *We know earth was peppered with thousands of meteors all at one time.*
- *We know that the meteors struck 10 to 12 thousand years ago.*
- *We know that the water temperature in the Atlantic and Antarctica both rose at that same time period.*
- *We know that the Water level of the Atlantic ocean rose by a huge amount at this time.*
- *We know that an Ice Age ended at this same time.*
- *We know that the earth's axis shifted at this time.*
- *We know that Venus's surface caught on fire at this time.*
- *We know that a large quantity of huge craters are located along Venus's equator and nowhere else as is the meteor that struck was rotating around the planet.*
- *We know that hundreds of written texts indicate that Venus changed significantly in recent past.*
- *We know that there is a plasma string between Venus and Earth indicating that some electrical connection was strong in recent past.*
- *We have heard that great commerce center Island nations sank in the oceans about this same time.*
- *We know that this was not the worldwide flood addressed in the Bible.*

I know you haven't been told about this previously, but it happened just the same. Here are a few of the many pieces of unimpeachable evidence.

**Worldwide Meteorite Evidence**-Large amounts of "meteoritic mass" and **an estimated 500 thousand strange indentions**, strongly believed to be from massive meteorite showers have been found around the world that date to the end of the Pleistocene era, about **10 to 11 thousand years ago**. Large quantities have been found in

<u>United States East Coast, Alaska, Siberia, Bolivia, and Netherlands</u>

Guess what! The time period for the destruction of the Venusian moon is about 11 thousand years ago. If they both happened about the same time, there is a good possibility that they were the same event.

**Glass & Stone Evidence**

Tektites are small pieces of glass formed as a meteor strikes the ground and melts the surrounding area. Many have been found in sort of an "S" shape and distributed over large portions of the Earth. *Some were found embedded in fossilized wood, in Australia, others were found in Vietnam, others were found in the Indian Ocean.*

Several dating methods were used including Stratographic, Carbon 14 and others. They showed that most of these pieces were **deposited around 10 thousand years ago**. Ok! Maybe the ones inside the fossilized wood came from an earlier strike, but most were Pleistocene Era events just like the Venus moon blast.

**New Zealand Evidence**-Today, huge quantities of metallic meteorites as well as objects called "china stones" can be found everywhere on the island nation of New Zealand. Inside the stones are the remains of burned up **Pleistocene** type material, which dates the event to between **10 and 15 thousand years ago. [I suppose you think these came from the Venus moon strike just like me.]**

**Carolina Bays Evidence-**While the above examples show an unbelievable stress on the earth, the best examples of meteoric evidence can be found in the United States. The east coast of the United States was pelted with huge quantities of objects 12 thousand years ago. There are still an estimated **500 thousand meteorite indentions** called "Carolina Bays", which mark this incredible event in history.

---

*Let me wonder for a minute at an absurdity. One hundred and forty thousand of these 500,000 meteoric blast holes have diameters of over 250 feet and you probably have never heard of them before now.*

---

I'm not sure what is scarier, this terrible event or the fact that people EASILY ignore it simply because it makes them uncomfortable. Just think about how uncomfortable the people of that time were as they essentially saw the sky fall all around them. The picture below shows the major areas where these objects have been found in the United States. **These generally date around the same time. The evidence shows that the Venusian moon most likely met its end at the same time that these 500 thousand holes appeared and the other holes around the world described above.** Some of these indentions are very large and have diameters that are thousands of feet across. So it wasn't just a little meteorite storm.

The direction of the blast also gives us useful information about our developing planet. The clear indication tells us that the "11 thousand year ago equator" was in the direction of the meteorite path. Quantity of the huge Carolina Bay Craters [greater than 1000 feet across] are listed below.

| State | # holes |
|---|---|
| Georgia | 27 |
| South Carolina | 102 |
| North Carolina | 202 |
| Virginia | 17 |

The Carolina bay incident was a huge onslaught of meteors striking the Earth, which caused holes everywhere. Don't just take my word on this. Below is a picture showing the quantity of these things in a small area. To make them easier to see, I put rings around the larger ones. There are 25 or more in this area alone and they are found along almost the entire Eastern coast of the United States.

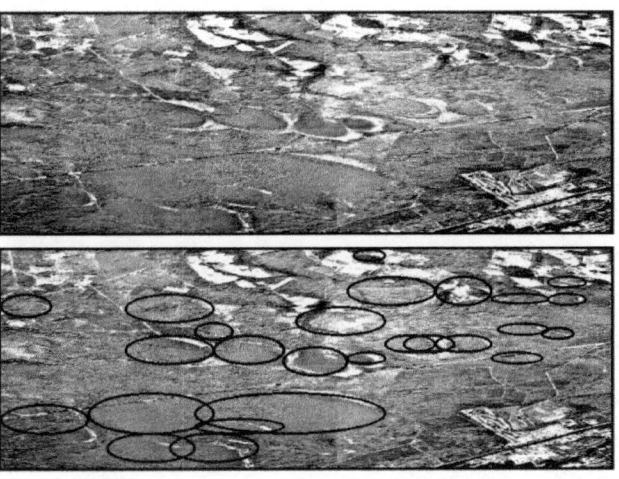

Just think about how it must have been that long time ago if you happened to live near South Carolina and literally hundreds of thousands of meteorites blasted the entire area over a period of perhaps 6 months. Most of your friends would have been killed and any civilization that was built up would have been in shambles. That is what the world would have been like, but the worst wasn't over—and for those who think that the Carolina Bays look like a multitude of sink holes caused by eroding caverns in an underground aquifer, think again. The underground area below these "Indentions" is not a limestone honeycomb and **sink** holes could not be the cause of the thousands and thousands of craters. These came from meteors and the meteors all came about the same time [near the end of the Pleistocene]. The explosion that caused the meteors had to be fairly close astronomically speaking because the pattern is fairly confined along the path shown.

This whole Carolina Bay thing is so revealing that it should be studied in our schools, because it shows us evidence of the earth shifting on its axis sometime after the bays were formed. It also establishes the cause for extinction at this time. It makes us look a Venus more closely.

# Venus Physical Evidence

hundreds of craters around the world is one thing, but if we are to believe that Venus is the actual planet that caused wide destruction of the Carolina Bays, Venus would have had to have been closer to the earth. After the incident, Venus would have moved away from the Earth and have taken its new position in the solar system. It would have been similar to when Earth moved away from Mars 300 thousand years ago as discussed previously only this time the movements would have occurred only 11 thousand years ago. With the Earth and Venus being close together, strong interactions could have come into play that eventually disrupted the "weaker" planet. Remember, we already have two major elements of evidence. One is a large set of accounts from all over the world that directly indicate that huge meteors fell from Venus. The second is the fact that the surface of Venus erupted into an inferno very recently. Now scientists have obtained a third piece of evidence, which shows that Earth was involved in the destruction. The evidence is called "plasma trails".

## Venusian Plasma Evidence

Outer space and our planet are both filled with something called plasma. Plasma is just a name given to a connected group of ionized atoms that can be influenced by magnetic fields. This plasma is strange stuff. It can even conduct electrical current. This current flow creates a magnetic field, which also affects the structure of the plasma. Scientists have not been able to adequately model this plasma phenomena mathematically, but we have a great deal of empirical knowledge of plasmas, because we use plasmas every day. Arc lamps, arc welding, and even a neon light are all basically plasma generators. Plasma trails can be and are produced from violent electrical disturbances occurring on planets, if they are significant enough to cause **quick atomic ionization**. This atomic ionization would be expected if parts of the planet were yanked away or hit with a huge electric field or hit by a huge lightning bolt. Additionally, we must understand that plasmas, although they are basically lumps of gas, do not behave like gases. They develop structure. When a huge variance in electric potential gets too high it produces a huge electric current that, in turn, causes the ionization. The current flow also causes these huge magnetic fields that, finally makes the something we call plasma filaments [**tails**] that twist together into things that look like "gas ropes". As long as the current continues, the structure of the plasma remains intact. Sometimes these ropes become very visible. If Earth was affecting Venus, there would be plasma tails between earth and Venus. So let's go to Kohistan.

**Ancients Saw Lines of Venus**-In Kohistan, there is a cave full of cave drawings. Researchers insist that a planet and star pattern depicted shows the alignments of stars as they were over 10 thousand years ago. Here is the thing that I particularly like--- there are lines drawn between Venus and Earth. They look like Plasma tails between the two planets. While this is evidence that these plasma things were around about  the same time as the Carolina Bays, we have found that the plasma tails were not visible only in these ancient times.

**Plasma still extends from Venus towards Earth**-In mid-1997, the Soho satellite detected a plasma structure issuing from Venus and it is long enough and in the right direction to almost reach the surface of Earth. The report described the structure as "stringy." Such a structure could only remain intact if a current were continuously flowing from Venus to the surrounding space via the plasma tail. Some researchers believe that the initiator could have been uneven electrical charges between Venus and Earth. No matter what initiated it, there is a high probability that pieces of Venus's moon were split away during the initial ionizing blast. These pieces would have fallen on Earth as a giant meteor storm. The discovery supports the idea that Venus assumed its present position in the solar system only recently, and has not yet achieved charge-equilibrium with its environment. When I say recently here I mean less than 40 thousand years.

The findings also give evidence to the probability that the reactive partner in the production of the plasma was Earth. It does another important thing as well. It makes the ancient descriptions of Venus even more believable as the planet would have looked like a huge comet in ancient times when the plasma trail was at its greatest size. It would have been a "wavy haired" planet and substantially more visible that it is today.

**Venus Looked Like a Comet**-Besides the wavy haired planet, various ancient names of Venus including Long Haired Star and Bearded Star, along with the other descriptors which typically symbolize comets, sound like very strange ways to describe Venus now, but that would have been exactly what would have been witnessed in the past if a plasma trail was visible at the time. In the early days, the electrical connection between Earth and Venus was possibly very pronounced and the planet was most likely closer than it is today. With these factors, that plasma tail we just discussed would have become visible and would glow just like a gigantic comet.

**Venusian Heat Evidence**-Today, Venus has a surface temperature of 900 - 1000 degrees F. and scientists are trying, unsuccessfully, to explain the extremely high temperatures away with a "greenhouse theory" that doesn't work. The planetary surface is so new that even the mainstream scientists are now having to devise a "global resurfacing event" [like the one presented] to explain it." We should look at all the similar myths and legends around the world describing a world-destroying catastrophe with Venus as causal agent and open up to the possibility that this well documented event could have caused havoc on Venus. Besides its lack of charge-equilibrium, Venus is totally out of "thermal balance" according to all direct observations.

- *There can be little doubt* of the following reasons for Venus getting very hot:
- *Venus probably was closer* to the Earth in the relatively recent past.
- *Many characteristic,* in-line, *identically sized, blast craters point to a war on the planet. This war might have been instrumental in the eventual doom of the planet.*
- *Some huge electrical disturbance* left characteristic lightning bolt marks on its surface and built plasma tails. The culprit was earth or something on the earth.
- *Some event apparently caused* Venus's moon to explode.
- *Many of the pieces hit the Earth* and Venus along both planets equatorial regions. The tight band of craters on the earth suggests that the source of the meteors was extremely close.
- *The same event, most likely, moved Venus* slightly closer to the sun. Historical references acknowledge this and the burning atmosphere enhances the possibility.

All these things in combination initiated the thermal changes we are witnessing today. It was not the mysterious "greenhouse thing" so go ahead and use underarm spray if you want to. It's not going to destroy the earth's Ozone and push us into a cataclysmic greenhouse melt down as TAUGHT IN OUR SCHOOLS. The earth is not going to burst into flames.

**Backward Spin Evidence-**In order to gain some sense about its thermal problems we need to not only look at the ancient histories. or the thousands of meteoric indentions, or even the in-line blast marks on the Venusian surface. Let's look at the fact that Venus spins backwards. This would not have been primordial as the planetary spins would have taken on the general forces that were found during the solar system beginnings. Most of the planets follow this rule, but the Venusian spin was changed by some traumatic event much later. The most logical event maker would have been interaction of a close planet [possibly earth].

166

**Venus Crater Evidence**-Besides the fact that there are very few craters on the surface of Venus which shows that the surface is very "young", we come to another curiosity. While it apparently makes no sense, almost all the craters can be found between 78 and 85 degrees in Latitude, but they can be found all the way around the planet along this central hub. Let's think about this for a minute and as we think let's look at Martian cratering.

**Martian Blast Difference**-If you remember from a previous section, Mars has extremely unusual cratering with almost all of the meteoritic craters located on one hemisphere. I showed that this crater grouping was not from a blast, but instead was the remains of the "old Martian surface". The other half of the planet had been ripped away 200 million years ago. The pieces of Mars that had been blasted along with pieces of the earth that had been pulled out into outer space must be still in orbit as they would have the same centrifugal forces as the planets. They would have made a ring of planetoids and someone could call the things an asteroid belt if they wanted to.

**Venusian Cratering**-On Venus we have a completely different phenomenon. This ring of craters tells us that the meteors came from something orbiting <u>very close</u> to the planet around its equator.

*I'll tell you what I think was orbiting Venus, but you have to promise not to tell anyone else, because they will think you are nuts. I think it was a moon.*

Now that I've said it, I know you think I'm ready for the loony-bin, but before I go, please look at the table below It shows the significant craters. Besides the fact that the crater density on the planet is fairly low notice that almost all of the craters are located around the equatorial region of the planet, just like I said. The ring of meteors is along the equator and varies only by +7/- 2 degrees. Moons revolve around a planets equator and if one exploded, it would make a line of craters around the equator.

| Name of crater | Lat. | Long. | Dia [KM] |
|---|---|---|---|
| Janice | 87 | 262 | 10 |
| Hua Mulan | 87 | 338 | 24 |
| Tatyana | 85 | 212 | 19 |
| Landowska | 85 | 74 | 33 |
| Ruslanova | 84 | 17 | 44 |
| Sveta | 82 | 273 | 21 |
| X | 82 | 85 | 12 |
| x | 82 | 85 | 2 |
| Odilia | 81 | 200 | 21 |
| Lagerlof | 81 | 285 | 56 |
| Efimova | 81 | 223 | 27 |
| Tursunoy | 81 | 229 | 5 |
| Eugenia | 81 | 105 | 6 |
| x | 80 | 229 | 6 |
| X | 79 | 270 | 3 |
| Nuon | 79 | 337 | 7 |
| Rudneva | 78 | 175 | 30 |
| Dashkova | 78 | 306 | 45 |
| Gina | 78 | 77 | 15 |
| Klenova | 78 | 104 | 141 |

## Weird Isn't it!!

I know you have never heard about Venus having a moon and you haven't heard about the disaster that caused it to become superheated happening only 12 thousand years ago, but scientists have discovered the evidence including the stuff I previously told you, the fact that all features on the Venusian surface are "New" features, a huge split along the axis of the planet, and even the burning temperatures all point to the same thing. Twelve thousand years ago Venus had a bad time and the earth was greatly affected as well.

**Venus Split Evidence**-Did I say "huge Split?? Yes I did. The photograph following shows the incredible split mark across the surface of Venus.

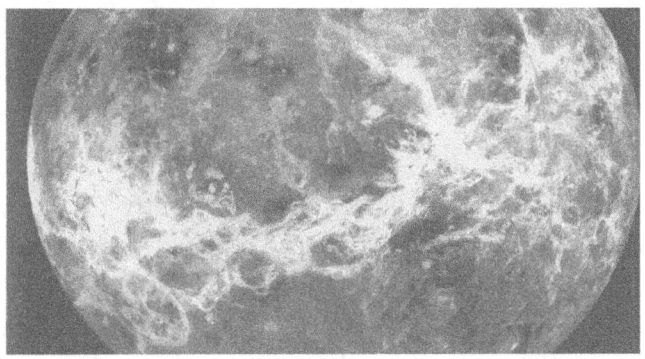

I know you are probably thinking that this picture must be a fake, because no one had  EVER told you about this calamity. You are probably thinking that if there was a huge gash across the surface of Venus, it would have been headline news. Some might think that the bright area is simply a photographic anomaly, but it is not. Let me show you one of NASA' topographically highlighted detail of the area so that you will recognize that the streaks are not some photographic "ghost". The planet almost split in two from the looks of the extremely long fissures. When I say long I mean a gash that is tens of thousands of miles long.

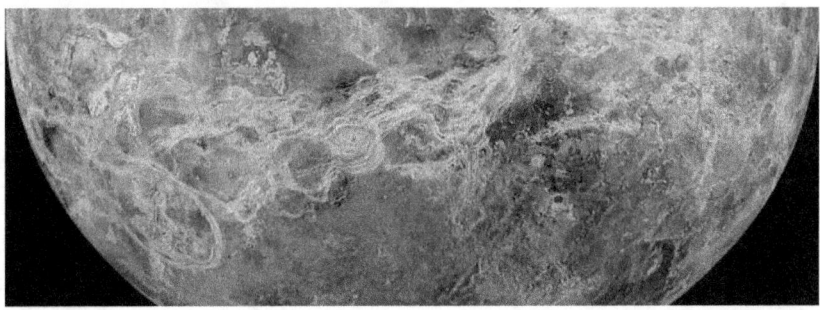

Note the fairly thin lined fissures along the center of the photograph of Venus. This is not a river. It is not a lava flow. It is not some "normal" characteristic one might find on a planet. It is the undeniable splitting of the crustal surface of the planet. VENUS almost split it TWO!

**Argon Evidence-** To date the catastrophe on Venus, scientists use Argon. A curiosity was found by the Magellan probe The curiosity was that the atmosphere contains high levels of the isotopes of argon, neon and noble gases. These high concentrations of noble gases <u>could only mean that the current</u> <u>atmosphere of Venus is extremely young</u>, because noble gases don't combine with other materials and escape easily into space; even with a thick atmosphere.

**Venus Habitation-**Here is something many researchers stay away from because of its present temperature, but evidence suggests that Venus was more inhabitable than Mars during the very ancient times. Because of an enormous amount of evidence we now can be almost certain that Venus was not always this ball of fire. In fact the flame out occurred fairly recently and many of the present surface features are extremely young. Even without its now heavy atmosphere, the climate would have been very warm during ancient times, but Venus would, most likely, have been green and beautiful. Maybe a little too warm for us to comfortably live without air-conditioning, but plants must have thrived and been in abundance. Many people, most likely, lived there as well because of the relative closeness to earth and its lush landscape. Some kind of catastrophe started to cause many volcanoes to erupt and the life of the planet began to choke. The atmosphere did not simply get thicker and thicker and finally change into an inferno as is presented in the "Greenhouse Effect Model". Whatever it was, the Venusian surface began to come alive with volcanic action from the trauma.

**Venusian Landscape Evidence-**Heat was noticed on both planets. Almost everything melted away on Venus, but not all evidence of Venus's previous civilization has been lost. Venus has its own, apparently man-made features similar to those found on Mars, but they are certainly not as well defined as those found on Mars simply because of the high temperatures and thick atmosphere. The picture [following left] is only one of the many scenic pictures taken by the Magellan space probe. Note the flowing river meandering down from the hills just like some beautiful spot on earth, but now the river is dried up and the vegetation is all gone. Before the atmosphere began to burn everything, the land must have been beautiful. We can see similarity with earth rivers. Whoever lived near the awesome river above died along with everyone else.

Everywhere you look are the remains of mighty rivers that covered much of the land. Venus was truly a beautiful place at one time. It is easy to see a difference between riverbeds and lava flows or other non-water related characteristics. The meandering nature and continuously similar sized flow points only to water. Because the rivers are still recognizable, the loss of the water happened very recently.

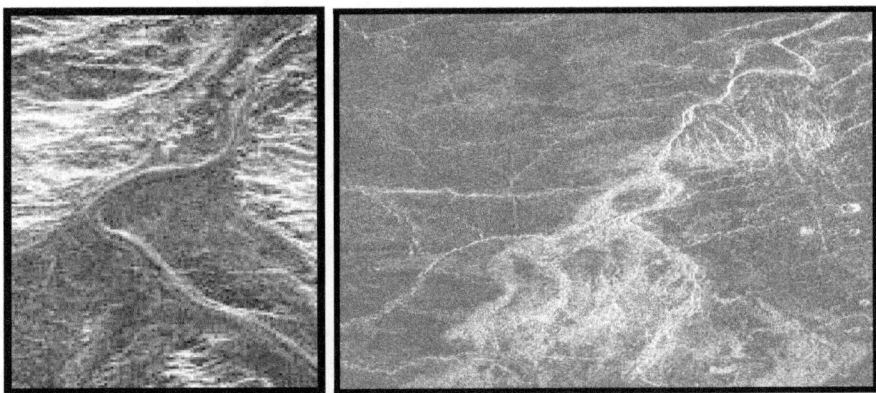

Above right is another one of the dried up river deltas found. This quick dry time could not have been hundreds of thousands of years ago or the details of the rivulets would have been consumed by the planetary actions, erosion, liquefaction and other tectonic occurrences.

# Atlantis and Venus

Everyone on Venus was gone and something mysterious started happening on earth. If having meteors spattering the countryside wasn't enough, the earth began to sink. All those who have heard about Plato's Atlantis descriptions raise your hands. Not you!-- you'll lose our place. I'm not going to go into this part of earth's history too much because a book could be written just on this single thing. In fact, many books already have been published containing various versions of the details of this horrible event. Anyway, Plato described an Island nation that was one of the central hubs of commerce just before the Venusian explosion time period. Other texts confirm his story and include as many as 6 other commerce centers at different locations. Over a period of years, the water level rose and rose until the inhabitants were forced to move away. The island nations went into the ocean never to rise again. Many of the survivors  went to Greece while others settled in Egypt. The story sounds surreal but the timing puts it during the time of a great upheaval on the earth, so we should investigate its probability.

In order to make the story true, the  of the things that caused Atlantis to drift under the water must  be found. There is a high probability that it was associated with the event that caused Venus to catch on fire. No! it was not underarm spray causing a greenhouse effect like the greenhouse fear mongers exclaim.. It was the huge explosion that almost ripped Venus in half leaving a gash that is about 25% around the surface of the planet. Whatever exploded near Venus and forced the Venusian disaster also caused the earth to be peppered with THOUSANDS of meteors along its equator and islands might have sunk.

## Catastrophic Elements

- *For maybe a thousand years, life on earth would be more unbearable than you could imaging.*
- *The meteoritic material killed thousands*
- *The fires that followed killed thousands*
- *The sudden change in the earth's axis killed thousands*
- *More than likely the sun was not bright during this time because the dust particles from the meteors and plants died off. This made a terrible drought and famine which killed thousands.*
- *Several ancient texts indicate that 1/3 of the population was destroyed at this same time.*
- *Something made the ocean depths to begin increasing which killed thousands and made many more homeless.*

Oh yeah! ALL of the evidence indicates that this happened about 10 or 11 thousand years ago which is about the same time as the Atlantis sinking. While everything above sounds bad the oddest is the last thing. Is there evidence that the water level did, in fact, rise, and where did the water come from? Below is some of the available evidence to consider.

## Water Height Evidence

The water level increased during this time period, but it took a very long time. The water level got higher each year over a period of thousands of years. Only after a long sinking time did Atlantis submerge. One could say that its submersion was because of something we call the Wisconsin Ice Age. To test this scientists have been testing the Atlantic Ocean. We can be fairly certain that the Atlantic Ocean is substantially higher on average than it was 12 thousand years because many have been checking and many cross referencing methods were used to test the data taken. As shown below: the average water level has increased from between 100 and 200 feet over the Atlantic Ocean's maximum height 10 or 11 thousand years. The data below is not just from one study, but is a consolidation of over 24 major studies. They all say the same thing.

| Testing Method | # of studies | depth inc. |
|---|---|---|
| Isotope-oxygen data [on the volume of seabed sediments] | 5 | 109 feet |
| Calculation on the basis of gravitation anomalies | 5 | 133 feet |
| Paleo-glaciological data [on the amount of the glaciation] | 7 | 126 feet |
| Geomorphological data [on the ancient coastal features] | 6 | 112 feet |
| Calculation on the basis of isostatic effect | 1 | 167 feet |

Prior to 12 thousand years ago, the water height fluctuated by another 200 feet as well. The graph below shows the relative water height of the Atlantic Ocean from 40 thousand years ago until today. Like the references above, the data was taken from many different studies. It shows the rising ocean described above. You will note from the table that the water level did in fact rise over a long period from about 15 thousand years ago until 10 thousand years ago. The rise is dramatic at over 400 feet during that time period, but was there an island that could have been the Atlantis of Plato?

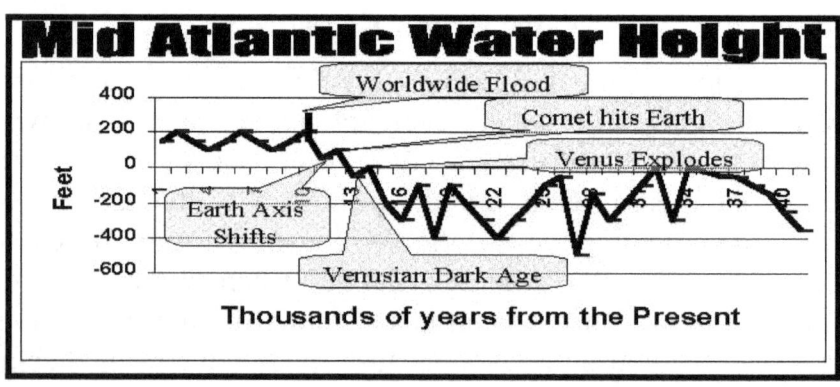

174

## Island Evidence

The drawing below shows what the world looked like with the water level 400 feet below what it now is. A few things that pop up are that the Azores becomes a huge island in the middle of the Atlantic Ocean, the Red Sea became a river, there is a huge island in the Indian Ocean, and the Mediterranean was simply a group of huge lakes connected by rivers. Anyway, there could very well have been a huge island nation right where Plato told us it was.

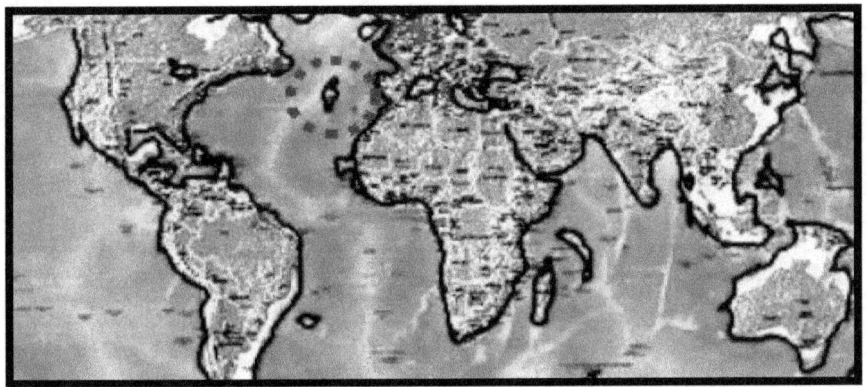

# Why did Atlantis Sink?

### The Water Came From Ice

From the table below, the late Wisconsin Ice Age is timed right for the Atlantis sinking event. With the beginning and end date generally exhibiting the highest water levels and the mid-Ice Age time period exhibiting the lowest water level. The golden period of the Atlantean reign would have been 18 to 13 thousand years ago. Then the water slowly rose over the cities and the island was lost as the ice melted.

| Name | Started [x000 Yrs ago] | Ended [x000 Yrs ago] |
|---|---|---|
| Illinoian/ Great Ice Age | 100 | 80 |
| Early Wisconsin Ice Age | 65 | 59 |
| Late Wisconsin Ice Age | 20 | 11 |

The picture following shows the amount of ice that was believed to have been covering the world during the last Ice Age. Much of it was converted into water to drown Atlantis. One thing to notice about this map is that there is an apparent shift in the ice edges from the current spin axis of about 30 degrees. The other thing to notice is that with all of this ice, many now covered islands would be above the water line.

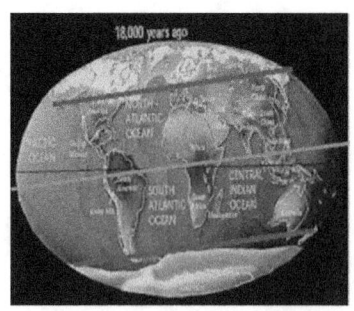

I know that the apparent equatorial line is not that indicated for the Carolina Bays so don't give me a hard time. From the table above, the Ice age started about 20 thousand years ago. It is reasonable to believe that at that time the earth axis was as shown and moved into the position indicated by the Carolina bays by the time of the Great Venusian Disaster. Also you might have noticed that the water level began to get higher about 15 thousand years ago which was before the Venus thing.

Let's examine this Ice Age thing for a minute. Although I do believe that the earth was colder some of the ancient time period, the whole glacier thing should be looked at a little more closely.

### The Glacier Error

*Have you wondered how the Glaciers could have moved thousands of miles <u>uphill</u> as commonly attested to by "glacier experts"?*

The well-known investigator, Herbie Brennan, proposed that water going uphill just has to be wrong. Just think about the supposed onslaught of Glaciers that covered much of Europe and the United States during portions of our history. Many times the only way for the glaciers to have gone where they did was this idea that glacier water could go uphill many hundreds of feet. Physics has a hard time with the concept and so should you.

The reason people think glaciers did this impossible task is that rocks were left on top of mountains and in valleys that have been scraped before being deposited in unusual places. These rocks are called erratics by the glacier scientists. Unfortunately, these same erratics have been found in the Sahara Desert, in Uraguay, and in Mongolia even though no one believes that Glaciers passed over those areas. For someone to present massive glacier movements covering the northern part of our planet as a fact, seems inappropriate. Don't through one science away to support another science. They need to work together. As part of the glacier thing, Mammoths were depicted lumbering through thick glacier fields. [No Way!!]

## The Mammoth Error

Have you ever wondered why they always show Mammoths in snowy regions? Certainly, they had 20 inch long fur just like the long fur associated with orangutans and the huge manes of lions, but that doesn't mean that the animals lived in snow. Elephants of all types <u>need huge quantities of vegetation to survive</u>. This huge version must have had an insatiable appetite for flowers and such. Whenever remains of foods are found in mammoth carcasses, the foods are flowers and vegetation. They could not have lived in the cold regions of the world. Mammoth bodies have been found in Alaska and Siberia, but, "guess what?" those places must have been warm if mammoths were there. For someone to tell you otherwise is inappropriate.

## Axis Movement Error

Let's first start out with a very important statement.

---
*THE GLASIERS GENERALLY DID NOT MOVE.*
---

If the glaciers didn't move, you might be wondering just how the ice got there. The answer might be found in the earth's axis. I don't mean the shift in the crustal areas with respect to the earth's spin axis, I mean the spin axis of the earth with respect to the sun. As shown below, today our planet has a 23 degree tilt with respect to our revolution axis around the sun. The areas with the most variation with respect to amount of daylight are located at the poles. Sometimes these areas have 6 months of darkness and with the darkness comes bitter cold. A spin axis perpendicular to our revolution would make the mildest pole temperatures while more incline means more cold.

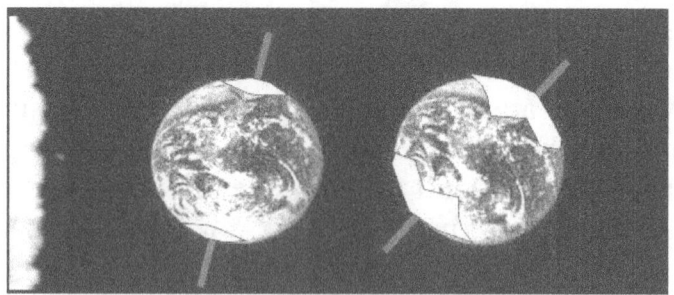

The 23 degrees is far from stable and has shifted many times. If the shift increases the incline we have an Ice Age. The shift that increased the water available and destroyed many island nations would have been a shift towards a decrease incline. This is much more in line with physics than the water going uphill thing and we can date just when the earth's axis became less declined and has generally continued its 23 degree angle.

## Another Possibility

The other possibility, I mentioned before. The earth spin axis with respect to the sun might not have changed, but instead, the earth's orbit could have been farther away up until about 10 thousand years ago. Whether the shift was caused from the interaction between Venus and Earth or the earth shift came first and began the whole earth Venus thing, is not known, but any of these things are more probable than glaciers moving uphill as you have been taught.

## City Evidence

There is evidence that this first sinking was slow and did not cause the massive tidal waves and upheaval felt during the "worldwide flood" that we will study later in this book. The relative slowness of the rising waters allowed some of the artifacts from before this flood to remain. Some of the remains of the magnificent preflood cities can still be found today. We don't find one city, we find a dozen. Unfortunately, you have to look under the water. One of these many undersea cities could have been the famed Atlantis. Surely that wasn't its real name, but Plato burned its name into our consciousness, so I will use that name to generalize the island nations of our past. Below are some city artifacts that have been found underwater. The evidence suggests that the majority of the destructions occurred around the end of the Pleistocene. Many cities, evidently, ended up underwater during this "special" time and the water never receded down to its preflood levels again. The fact that these cities were already submerged actually saved their remains while those cities above the water were almost completely destroyed in the huge tidal wave actions of the more famous worldwide flood to come.

***City Near Morocco-***Roads and buildings were found in 60 feet of water off the Moroccan coast.

***North Sea City-***In 1954 the remains of a city was found in 50 feet of water. Molded slabs of firestones for road pavement and smelted ore products were found. The estimated age of the city was at least 3500 years old, but it could have been much older. [**I say, 7 thousand years older.**]

***City off Coast of Spain-***Off the coast of Spain, the remains of a city that sank in the ocean thousands of years ago was found in 1973. According to *"United Press International"*, the remains included walls similar to that shown below left, large columns, and roads.

***Off the Japanese Coast-*** Huge underwater "temple-like" structures have been found. [See above right] Possibly they are the remains of a thriving city that flourished thousands of years ago. A huge temple structure and what appears to be a roadway are being investigated. The photo shows some of the regular "Steps".

***Portugal-***Four thousand miles off the coast, Russian scientists found buildings made of strong concrete and even plastics. Along with the buildings, they found the remains of city streets. They brought up a statue from the area for study.

***Black Sea Find-***Off the coast of Turkey, large blocks associated with a city that sank thousands of years ago have been found.

***Off the Coast of France-*** A flight of stairs was discovered in 1964 leading downward off the coast while researchers were in the bathescope Archimedes.

***Greece-*** While trying to find the arms of the statue Venus, Jim Thorne found an underwater citadel near Melos. The remains were several hundred feet deep with roads going even deeper.

***Malta Waters-*** Cart tracks going everywhere on the island also go straight into the Mediterranean Sea. The roads appear to connect the island with the mainland. This "center of the Mediterranean" Island was probably once part of the mainland and was part of a much larger civilization before the initial water rise.

***Rhode Island-*** In 1958, near the Brenton Reef, skin divers reported finding a conical shaped tower estimated to be 50 feet high in 90 feet of water. Each of the stones on the tower was estimated to be the size of an average desk.

***Peru-*** Photographs of carved columns were reportedly taken at a depth of 1000 fathoms off the coast of Peru. Sonar soundings also indicated possibility of buildings at the site.

***Bahamas-*** A pyramid, roads, columns, a dome, rectangular buildings, a statue, unidentified metal objects, and a 1600 foot long, underwater wall made of limestone blocks which are 15 to 20 foot square have all been found off shore. The picture below shows a section of the wall.

# Plato's Atlantis

Ok! There are many underwater cities, but we need to go back and first get a feel for what Atlantis was all about before we continue because we have heard about Atlantis as a myth so long that it sounds funny talking about such a place. I know this seems like a bizarre subject, but you have already gone through many bizarre subjects and you are still reading this history, so who's the oddball? Don't get me wrong. I haven't been trying to find strange elements of conjecture to push into some type of fairytale. I am bringing you a cross comparative collection of elements to support a probable history that just happens to link some of the stranger writings and physical evidence together with more mainstream science and religion. Atlantis fits this mold.

Everyone has heard about Plato's Atlantis description covered in the book "Timeaus", but there is one aspect of his history that should be brought out a little more. It deals with the dual flood concept which is important to understanding Earth's development.

### Timaeus 22d-23c

*And the reason is this. There have been and will be many different calamities to destroy mankind, the greatest of them by fire and water, lesser ones by countless other means. But in our temples we* **have preserved from earliest times a written record of any great or splendid achievement** *or notable event which has come to our ears whether it occurred in your part of the world or here or anywhere else; whereas with you and others, writing and the other necessities of civilization have only just been developed when the* **periodic scourge of the deluge descends, and spares none but the unlettered and uncultured,** *so that you have to* *begin again like children, in complete ignorance of what happened in our part of the world or in yours in early times...* [This indicates that the "Atlantis sinking" was not the last major catastrophe encountered by the Egyptians and Greeks.]

*You remember only one deluge, though there have been many, and you do not know that the finest and **best race of men that ever existed lived in your country**; you and your fellow citizens are descended from the few survivors that remained, but you know nothing about it because so many succeeding generations left no record in writing".* [The most obvious reason that the Greeks had no direct knowledge of the Atlantean refugees colonizing Greece would have been that another terrible flood or cataclysm occurred a long time after the Atlantis sinking. We have all heard of this second flood as it essentially covered the entire world a mere thousand years after the Atlanteans colonized Greece.]

## Plato's Flood Timing

Even Plato's timing matches the evidence and the presented timeline. I know you have been told that Plato's Atlantis sank 11 thousand years ago, but that is not what his history stated. Assuming the information was obtained about 1000BC and the Egyptians were instructed for 8000 years, and Greece was instructed 1000 years before that, then the time of the sinking was between 11 and 12 thousand years ago, which is the time period we are investigating. Let's read what was said exactly in both "Timaeus" and "Critias".

## Timaeus. 23d-24a

*"Solon was astonished at what he heard and eagerly begged the priests to describe to him in detail the doings of these citizens of the past. "I will gladly do so, Solon," replied the priest, "both for your sake and your city's, but chiefly in gratitude to the Goddess to whom it has fallen to bring up and educate both your country and ours - yours first, when she took over your seed from **Earth and Hephaestus**, ours **a thousand years later**. The age of our institutions is given in our sacred records as **eight thousand years**, and the citizens whose laws and whose finest achievement*

183

*I will now briefly describe to you therefore lived **nine thousand years ago**; we will go through their history in detail later on at leisure, when we can consult the records."*

## Critias

*"Let me begin by observing, first of all, that nine thousand was the sum of years which had passed since the war said to have taken place between those who dwell outside the pillars of Hercules and those who dwell within." **[With Plato living almost 3 thousand years ago, 9 thousand years from Plato was about 12 thousand years ago.]***

## Plato Wrote about a Second Flood

Not only did Plato talk about the flood that destroyed the Atlantean Island, he also told about the second more serious one. Let's continue reading.

## Timaeus 25c-d

*"At a later time there were **earthquakes** and **floods** of extraordinary violence, and in a single dreadful day and night all your fighting men were swallowed up by the earth, and the island of Atlantis was similarly swallowed up by the sea and vanished..." **[This was not referring to the slow increase of flood waters around Atlantis, but a separate and violent flood from which Noah survived according to Biblical accounts.]***

## Critias

*"Atlantis once had a greater extent than Libya and Asia and afterwards sunk by an earthquake." **[The description of sinking by earthquake rather than simple flood seems to go along with other data.]***

*"Many great deluges have taken place during the nine thousand years." **[Of course one worse than the Atlantean flood occurred later and practically everyone was drowned. We will find that this terrible flood occurred about 10 thousand years ago or about 3 thousand years after Atlantis disappeared into the sea.]***

# Ice Age Proof

One reason that the water level never came down to its pre "Atlantean flood" levels was "less ICE". If you recall there was something we call an ICE AGE about this time. I stated that Venus probably moved in closer to the sun around this time and that earth was affecting the planet during this time. The evidence suggests that earth was possibly pushed in closer to the sun a little just as Venus was being pushed toward the sun. The effect was obvious. As earth went closer, some of the Ice turned to water on the earth and stayed water until today.

# Evidence of a Shift

One thing that could cause a change in water height is a change in the earth's rotational axis. From several sources, we can gain a good picture about the earth shifting on its axis during the Venusian adventure. The texts tell us a gruesome tale about 1/3 of all life being killed during this struggle for earth to settle down.

### Enochian Evidence

In the book of Enoch, the event is described vividly along with the cause of the event. According to "Enoch", men had learned too much from entities known as watchers, and would not repent of their evil ways. A third of the animals and people all died due to this mistake.

*Enoch 67:14-"The water of the springs shall **again** undergo a change, and become frozen." [**This ancient writing tells of the drastic shift in the climate that may have come from the rotational axis shift. It also indicates that this type of shift had happened before. I know it also sounds like the cyclic Ice Ages of the Earth, but what causes those things?**]*

*Enoch 64:1-[**This ancient Jewish text confirms the event**] "In those days Noah saw that the __earth became inclined__ and that destruction approached. -The earth labors and is violently shaken. Surely I shall perish with it. -There was a great perturbation on the earth. [**The incline change denoted a shift and the perturbation denoted gigantic earthquakes. This verse was before the worldwide flood time.**]*

### Other Written Texts

We find the same story from Greek and Egyptian historians.

*Herodotus-According to ancient Egyptian historians and written down by the Greek historian Herodotus, "The sun did not always* *rising in the same place." [**The only way for that to happen is if the Earth axis shifts as would be evident from a crustal jump.**]*

***Egypt-****The Harris Papyrus states, "A cosmic cataclysm of fire and water followed with the* **south becoming the north** *and the earth turning around." [**This text dates the shift as being after the cosmic cataclysm of Venus**]*

### Physical Evidence of the Shift

Before you disregard the notion of the earth axis shifting from a multitude of meteors hitting it, here are a few of the many indications of the event.

**Straight Line Distribution Evidence**-I told you to remember the Carolina Bay direction presented earlier. Well, here is why it should have been noticed and remembered. If we look at the distribution pattern of the "Carolina Bays" it becomes apparent that there is a straight line of events that occurred around the world. The direction of the impact axis is shown below. In fact, if we follow the line around to the other side of the world, we find more evidence of massive meteorite strikes that occurred at the same time in **Australia**. You say SO WHAT!

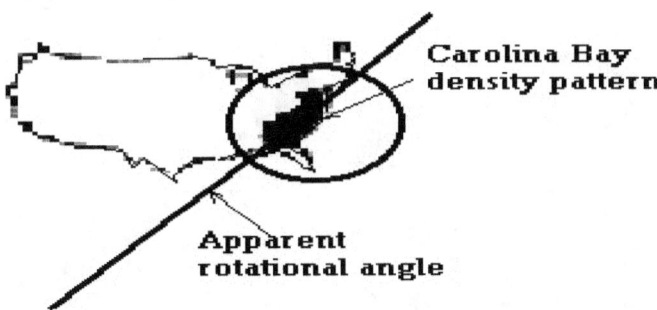

Besides giving us knowledge of a tremendous meteoritic event, the Carolina Bays provide us evidence of our last rotational shift on Earth and they even give us a good approximation of the previous axis of rotation for the Earth. That is because the density of the bays and the evidence in Australia show a "straight-line" distribution pattern that is consistent with a bombardment along the equatorial boundary.

If we consider the impact density line as the "Old" equator, a shift of about 30 degrees in the rotational axis has occurred since the bombardment as shown in the pictures below.

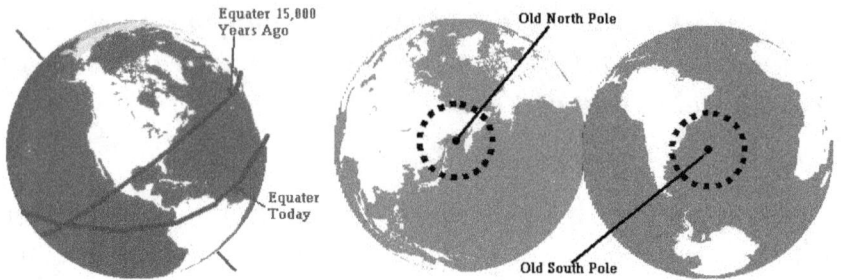

On the middle image, the globe has been separated at this ancient equator. Note that along the polar path there was not much land. Also note that eastern Alaska and Siberia are well away from the Arctic Circle, which allowed huge herds of Mammoths to dine on flowers in those areas just before the shift. The shift froze them solid. As we look at the shift across its equatorial cross-over point, we can quickly see that the Peruvian [preIncas] and Afghanistanian [Aryan] societies would have been thriving prior to the shift. They were both situated on the tropic of Cancer and Capricorn. After the shift they would have had to migrate.

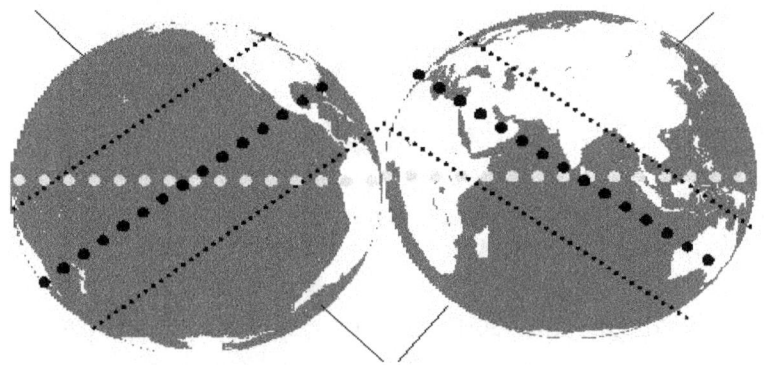

**Migration Evidence**

After the shift, the last shift the Afghanistan region moved north away from the comfort of the Tropic of Cancer and the garden spot moved into India as did a large segment of the world society. Peru moved into the Equator, so the Peruvian inhabitants moved north to Central America. The map following shows the general locations of major civilizations after the shift. The migrations shown all occurred about 10 thousand years ago.

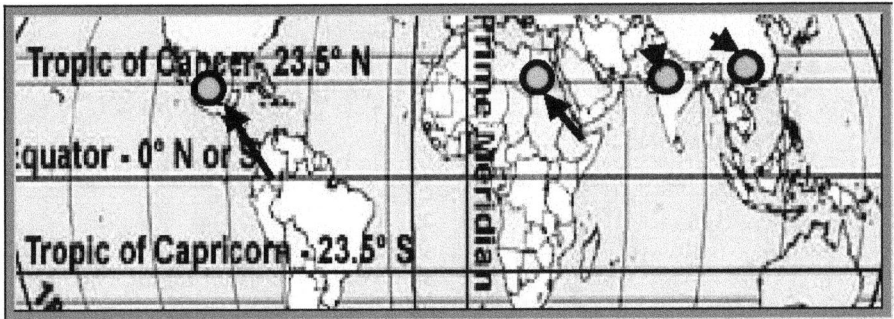

## Two Flood Evidence

While I'm talking about a great flood that overtook much of the lowlands of the earth, there is strong evidence that there were actually two major floodings that occurred within about a thousand years from one another. This first one was important in that many survivors were able to escape to the Americas, Europe and Egypt. Even though many did survive the first one, almost all the inhabitants were destroyed in the worldwide flood that followed fairly soon afterwards so don't get the two confused.

# More Timing Evidence

I know I've already stated that the particular flood we are discussing here occurred 11 thousand years ago, but I thought it would be nice to read some of the historical insights of the flood and the island nation. Plato made the city of Atlantis famous with his treatise about the huge civilization beyond the Pillars of Hercules which he called Atlantis. There are actually many references to Atlantis in ancient tests besides that of Plato. These references can be found from Egypt, India, Greece, China, Peru, Ethiopia, Persia and other places around the world. The evidence confirms the date and the incident. Here are some of the testimonies that let us know "when" Atlantis sank.

### Greek Timing Evidence

Two historians from Greece tell us the same thing. The historians were Plato and Strabo.

*Plato-From his third hand knowledge, Plato indicated that 9000 years before the time of Solon, or about **11 thousand years ago,** the city of Atlantis sank*

*Strabo-This Greek historian wrote that about **10,000BC** the Greeks fought the Tartessians [Atlantians] beyond the Pillars of Hercules which must have been before its' sinking.*

### Egyptian Timing Evidence

Evidence can be found in Egypt. Here are some excerpts from Mantho's works and several extant papyrus-es [or is it papyri?].

*Mantho-This Egyptian historian stated that 5800 years before, Menes, the first Egyptian king, was from the dynasty of "The Spirits of the Dead". [**The time would be about 10,000 BC. The dead probably is reference to the death of Atlantis.]***

*More Egyptian records-A papyrus in the museum of St. Petersburg stated "Land of Atlantis, whence had come the ancestors of the Egyptians 3350 years ago---the sages of Atlantis flourished during a period of 13,900 years"* **[Although the time of the Atlantis destruction cannot be determined from the 3350 year number, this does tell us that if the submersion was 11 thousand years ago, then Atlantis became powerful about 24 thousand years ago.]**

*Atlantis Time-line-In the "Manetone Papyrus" which can be found in the British museum, the Egyptian historian named Manetone provided some insight into the lineage of the kings of Atlantis. He indicated that the Atlantean kingship went back for 13,900 years before the initial beginning of the Egyptian kingship. King Thoth was the first king from Atlantis according to his own account in the "Emerald Texts". If we place Thoth at the worldwide Flood it would place the Atlantean kingship starting about 20 thousand years ago.*

*"Sent Papyrus" Evidence-According to the "Sent Papyrus", which is the oldest papyrus known to exist, and is resident in Petersburg Museum, in Russia, Pharaoh Sent dispatched an envoy to search for Atlantis from where the Egyptian forefathers had supposedly come to Egypt some 3500 years earlier bringing wisdom.*

### Tibetan Timing Evidence

*From Tibet, The "Book of Dzyan" says, "Large areas of land sank in **9564BC**."* **[I'm not sure why their date is so precise, but that would be 11,567 years ago.]**

### American Timing Evidence

This evidence can be found from the Maya, the Aztec and the Caribbean.

*Codex Troano-This Mayan text indicated that the destruction of **Azatlan**, by sinking, occurred 8,060 years before the writing or a little over **11,000 years ago.** It also indicated that millions of people died in the cataclysm.*

***Aztec History-****According to the Vatico Latin Codex the flood that sank **Azatlan** occurred **13,205 years ago.***

***United States-***Off shore excavations near the Atlantis site off the coast of the United States has dredged up calcareous disks about 15 cm in diameter and 4 cm thick with a round hole in the center. Radio carbon dating indicates that they were formed on dry land over **10 thousand years ago,** which provides still another level of confirmation of the catastrophe date.

## Pleistocene Timing Evidence

Geologic evidence concurs that much of Earth was destroyed by fire, volcanoes and ice about **11,000 BC**. This was near the end of the ***Pleistocene Age.*** Volcanic eruptions spewed everywhere and blocked the sun. This may have added to the coldness and the Ice areas, but it would have been too short lived to support low water levels for thousands of years.

## Seer Timing Evidence

*Edger Cayce had a vision about Atlantis. -His psychic readings indicated that before **10,500 BC** flood consumed Atlantis.[**OK! He could have read Plato's work, but this guy had many accurate visions of the future that have come true, so don't discount data just because it may at first seem dubious. Certainly recognize the source and temper its importance, but don't ignore date you don't like.]***

## Water Temperature Evidence

We don't have to simply rely on –
- *Dozens of water height studies,*
- *Dozens of ancient historical references,*
- *Physical evidence associated with the Carolina Bays,*
- *Suspect migration patterns,*

That certainly is not enough to make us think that the earth shifted, but what if we could show that the water got hotter all of a sudden and stayed hot. We can be fairly certain that the water temperature in the Atlantic Ocean abruptly increased as the Earth's axis shifted. The National Geophysical Data center provides the following information concerning the water temperature over the last 100 thousand years.

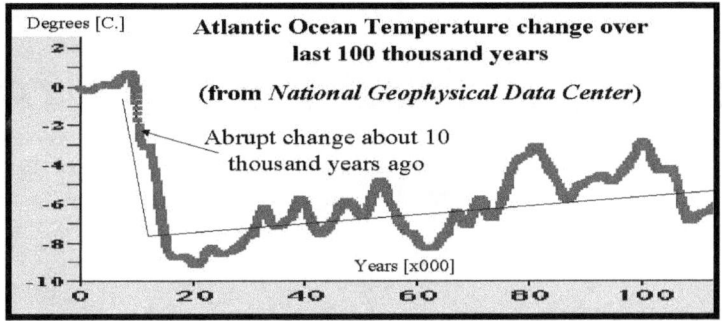

Note the change that occurred over about a 5 thousand year span from 15 thousand years ago until 10 thousand years ago. The average water temperature increased by 10 degrees Centigrade or 18 degrees Fahrenheit. One reasonable way for that change to occur is for the earth axis of spin to change bringing more of the Atlantic Ocean into the equatorial region as discussed above.

After the Venusian catastrophe and the changes that occurred on earth, there was a short respite. People began to get confident again and life became normal. For 3 thousand years life was ok again, but the world would see another drastic change a mere 9 thousand years ago. This time the entire world flooded, but that is another story.

# Biblical Destruction of Venus

For those who believe that nothing happened in the past unless the Bible said it happened, they will be happy to know that the Bible also identifies a surprisingly similar destruction of a planet called Rahab. It even gives an approximate date. Several verses talk about the power that was used to split it apart and there is a strong indication from the Biblical texts that pieces from the planet "split apart the sea". The name Rahab means "vain or vanity" in the Hebrew tongue and it is not difficult to imagine that a planet with flowing hair and was the brightest in the sky would have had that name. The Bible specifically identifies stones falling from the sky and the book of "Jasher" gives us an approximate date. The date was after the reign of Enosh. This would mean that the event happened 12 to 20 thousand years ago, by my expanded dating described earlier.

### Huge Destruction

"Jasher" also provides us with estimates of the destruction during the disaster. It indicated that 1/3 of the inhabitants of the entire Earth were destroyed. Here are the specific Biblical verses.

***Psalm 89:10 -*** *"Thou [God] hast <u>broken Rahab in pieces</u>, as one that is slain;"* ***[The pieces sound like meteoritic pieces. Especially as we read further.]***

***Isaiah 51:9-*** *"O arm of the LORD; awake, as in the ancient days, in the generations of old. Art thou not it that hath <u>split Rahab</u>, and wounded the dragon?"* ***[The Dragon most likely was one of the Nephilim people or a mighty creature governed by them.***

*Note the <u>idea that the planet was split</u> as is seen in the topographical map.]*

*Job 26:12-* *"The boastful Angel and his followers rebelled. Yahweh destroyed their dwelling places. He divideth the sea with his power, and by his discretion <u>he smashed Rahab</u>. It was reduced to **stones of fire**." [By this verse we could well believe that many of the Nephilim had made Venus their home before the disaster.]*

*Enoch 85 and Revelation 9-*"*I beheld a single star fell from heaven-then I beheld many stars which descended and projected themselves from heaven to where the first star was." [I know I brought up this verse before, but it could very well be the vision of many meteorites hitting the Earth.]*

*Jasher 2:5-6-* *"-and the sons of men forsook the Lord all the days of Enosh [Adam's grandson] and his children; and the anger of the Lord was kindled on account of their works and abominations which they did in the Earth. And the Lord caused the waters of the river Gihon to overwhelm them, and he destroyed and consumed them, **and he destroyed the third part of the Earth,** and notwithstanding this, the sons of men did not turn from their evil ways--" [We'll discuss how the 1/3 fits into everything shortly.]*

*Isaiah 14:12-* *How art thou fallen from heaven, O "Heylel" [**morning star**] , son of the morning! how art thou cut down to the ground, which didst weaken the nations! [I replaced the word Lucifer for the actual Hebrew- the usual translation of Heylel is Morning Star. Essentially, this is talking about parts of Venus falling to the ground and weakening the nations. The idea of weakening nations sounds like something hurting the human inhabitants of the earth such as one would expect from the huge meterorite storm aftermath of an exploded moon of the nearby planet Venus. By the way, this Heylel term is used no place in the Bible except for this verse.]*

I know some of this is sounding like a stretch, but what I have been trying to stress in this whole set of books is that if most of the pieces gathered from many areas seem to be painting a similar picture, then even the more improbable becomes the most likely if the pieces fit. Maybe there is a theory that can come from all of the evidence.

## Dropa Destruction Evidence

Before we go on, there is one more piece of evidence that should be addressed called Dropa. While not generally addressed here, it can be shown that the Dropa accidentally settled in China when their flying vehicle had trouble about 12 thousand years ago. They never left the earth. Their appearance was very different from the people lining in the surrounding area. You can look into this strange group of marooned people at many places on the internet and other sources. Their heads were large, their bodies were small and delicate, their eyes were large. I suggest here that one of the reasons they stayed was that there was no place to go home to. Their probable home was Venus and it had just had its moon explode and had begun to ignite into the inferno it is today. Below are believed to be 4 of the 716 "disk books" left by the group. These "books" are still being decoded today.

While I do believe people lived on our neighboring planets, I "DO NOT" believe in Martians and Venusians, per se. Both planets probably had "some inhabitants" in the ancient times, but the inhabitants were almost assuredly humans that had been sent there for trade or for war assignment. I don't believe that there is enough evidence to support the theory that either planet would have been more inhabitable than earth and no compelling evidence that the "citizens" came from anywhere else in the universe.

The bombardment of the Venusian moon particles didn't simply happen and afterwards things went back to normal on earth. The effects of the bombardment, evidently, set up a chain of events including pushing volcanic debris in the sky and initiating a heating cycle on the earth. Inhabitants were initially unaware of how catastrophic this would be, but soon their fate was known.

## Possible Structures on Venus

Even though the detail is substantially less clear through the thick atmosphere than images taken of the Martian surface, we can still see suspicious items that look man-made. The figures and related drawing, shown below, are just a few of the more suspicious items which identify civilization on Venus in the not too distant past.

*Item "A"-* appears to be some type of high walled building.

*Items "D" and "F"-* have the familiar triangular shape that most Venus artifacts have. Although triangular shapes are possible in nature, the regularity and equi-distance lengths make these items almost impossible to duplicate in nature; especially when considering the large quantity of buildings that have this same structure.

*Item "B"-* is usually called the arrow due to the wall directly in front of what looks very much like a triangular monument. Even details of three pillars can be made out and what appears to be steps.

*Item "C"-* not only looks like someone wrote the word HI but is typically called the Venus Stonehenge. 4 or 5 pillars are placed in a circular pattern and others lie on the ground in front of the pattern.

*Item "E"* is one of very few true pyramid structures found on the planet. These items and many more do not look like the result of natural phenomenon.

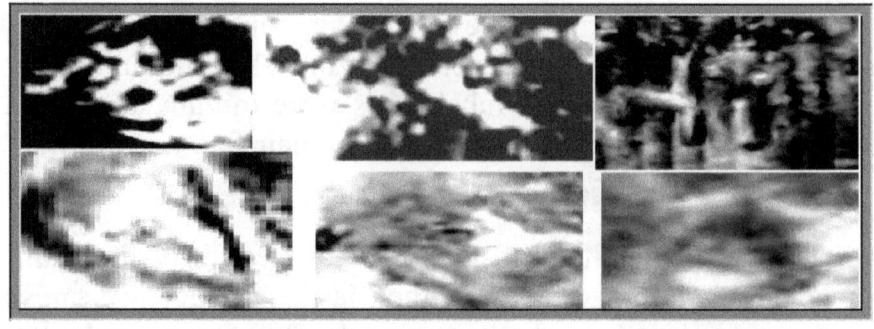

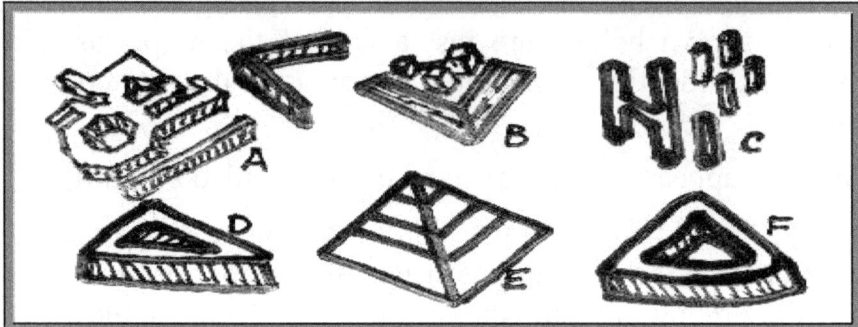

All but melted away, most of the evidence of its once thriving communities is completely gone. While it might seem that the Venusian communities were lost due to the terrible heat that now consumes the planet, that is not the case. The communities vanished when Venus almost split in half during the very recent times. Only after the great split [described earlier] did Venus catch on fire. Certainly the details of possible buildings and strange artifacts are limited and not as sharp as those being investigated on Mars, but coupled with other details of the war-on-Venus that can be determined from ancient texts, the probability of life on Venus at one time should be considered to be high.

## No Greenhouse

Scientists have been trying to attribute the change in atmosphere to something they call the greenhouse effect, but the timing just doesn't fit, in fact, the whole theory seems extremely flawed. The surface volcanic features are all relatively recent and there are very few impact craters which could have been the initiation means for such a catastrophic event. By the way, it was not Venusians using fluorocarbon deodorant spray that caused it to ignite, nor should you believe that the earth will have its own "greenhouse catastrophe" if they are used on earth. That type of mumbo jumbo is only good for businesses that make non-fluorocarbon deodorants.

## What Happened on Venus?

The planet did "Greenhouse", all right, but the reasons were not from events that occurred millions of years ago, the evidence suggests it happened fairly recently. Part of the evidence comes from the Magellan space probe which photographed 98% of the surface of Venus. All the large craters were found to be along the equator of the planet, this shows that the planet once had a "moon" and it exploded. This seemingly outlandish remark isn't what you have been told because the evidence doesn't match most of the theories. Therefore, the evidence was cast aside in favor of the disproven theories of greenhouse explosions for no reason beyond too much flouro-carbons being sprayed as underarm deodorant sprays or leaks for air-conditioners.

## Craters that aren't Meteor Craters

The row of craters below left are not indicative of a meteor shower that would cause a random layout of variably sized blasts as the meteor exploded high in the atmosphere. These strikes are directed in a line on Venus. Here are seven blast areas in line. Someone was apparently trying to hit something during a strafing run of some kind. One thing that should be noted is that each of the blast areas is exactly the same size so the blasts could not have been random pieces of meteor unless each piece came from the same source, all happened at the same times and all pieces were the exactly the same size and density. This is indicative of bomb blasts.

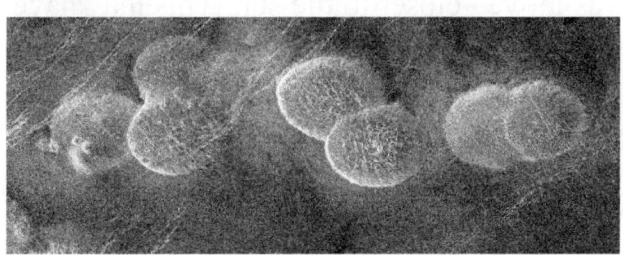 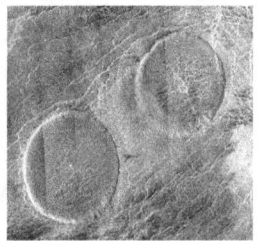

## Volcanoes that aren't Volcanoes

The preceding right image is a pair of perfectly round objects identified as volcanoes on Venus. Has anyone ever seen a volcano as accurately cut and with two identical ones side by side before? I don't think so. I don't know exactly what they are, but at least now you are aware of them. While there is an extremely high probability that Venus supported a substantial population in the early days it is certainly not the only one. Our next closest neighbor, Mars, was not always as lifeless as it is today. Before it began its loss of it atmosphere and surface water. It too most likely held life.

## Possible Timeline for Venus Development

The list below shows potential timeline for the development and demise of the civilizations on Venus. The number represents number of years ago.

**4.5 Billion**-The planet was formed

**200,000**-Due to its proximity, it is likely that Venusian Cities were established by this time.

**120,000**-Whatever slowed the earth's rotation down may have affected Venus as well. It is assumed that Venus had a somewhat similar rotation to earth during this time period.

**100,000** -While Mars began losing its atmosphere, Venus enjoyed a pleasant time. The whole planet would have been green at this time just like earth.

**25 to 11,000** -Several wars evidently involved Venus and earth.

**11 000**- A red letter day in the memory of Venus, Very few people got away, those who left the planet came back to earth, the planet's moon began to disintegrate. It is not known what caused it, but the earth had something to do with it.

**10,000**-The planet was moved closer to the sun. This may have occurred when the moon blew up or slightly after that time, but soon the planet was much farther from the earth than it had been; the planet began to slow down and essentially caught on fire. [This was the Pleistocene extinction on Earth and the end of everything on Venus.

## Scientists Breed Fear

I discussed the various planets and how there was some possible interaction between the planets and the people of the earth which may or may not have greatly affected the individual planets. I showed how some major encounter between earth and Venus initiated the fireball that has been consuming Venus over the past 11 thousand years. The thing I never talked about is the "greenhouse affect" and how "global warming" will eventually take over the planet and cause it to burst into flames. The reason is that the whole concept of global warming is a lie.

# The Global Warming Lie

Remember the halting of all "flouro-carbon sprays" and the testing of "methane producing cattle flatulence" because of the fear of the greenhouse effect? I'll bet you thought that there was "Global Warming" all over, because everyone kept pushing it into your head, but you have been lied to. This wasn't on purpose, necessarily, but it shows a major problem. The science community gets so wrapped up in a theory, that they will do almost anything to show that the theory works; **ESPECIALLY IGNORING EVIDENCE THAT GOES AGAINST THE POPULAR THEORY**.

## $CO_2$ levels

As it turns out, greenhouse gases are not higher. The opposite is occurring and has been occurring for many, many years. Scientists are now checking the $CO_2$ levels and they are disturbed because there was essentially no $CO_2$ buildup in the 1990s. That means that greenhouse gasses are lower than in olden times. According NASA satellite data and several Antarctic studies released in 2002, the entire Earth has really been **cooling for the last 18 years;** the western ice sheet is growing every year; and the interior valleys have been getting colder for the last **30 years**.

There is a lot of embarrassment going around for the hard headed scientists. Every year, scientists made up more excuses that still showed that the Earth was warming because we would not stop burning things, but the whole thing was a lie, possibly to keep us from burning things. **[Here is where there is a lot of conflicting information. Some scientists still hold on to the warming theory and papers are still being written about it every year.**

**If they write enough papers the earth will get warmer, but they will have to write very fast to cause enough friction.]**

Some of the data does suggest a very slight warming cycle, but more of it does not go along with this theory and there is NO data that suggests that anything we do about it has any affect. In all the millions of years of this cyclic behavior, the earth has **never** gone into a significant "Global warming fiasco". It's like saying Mars could, all of a sudden, get hot. It won't happen. Our planet has a hard enough time keeping itself warm for us. Remove the atmosphere or disrupt it in any way and the temperatures will plummet. You have seen information in newspapers and articles about the earth warming, but what if we are going the other way? It makes more sense and we need to wake up. What if I told you just this year, after 30 years, scientists had to move their entire research center as it was getting too far away from the open Ocean as massive amounts of snow and ice keep collecting year after year..

### Next Ice Age

Most likely, what is really happening, is just as bad as global warming, because we are evidently going steadily towards our next Ice Age. Many climate experts believe we are overdue for an Ice Age so they check for indications continuously. The ice core studies and long term weather research all indicate that the Earth alternates between ice ages and inter-glacials.

In 2002, at the South Pole, the penguins were stranded because of an abnormal ice buildup not because of the blazing heat the greenhouse model has suggested, but instead the weather was colder. This is only one of many indicators of the impending problem.

The graph following is the deuterium concentration at different depths of ice. There are two things to notice. The first is that since about 11 thousand years ago, Antarctica has been getting colder. The slope represents about a 12 degree drop in average temperature over that time period. The only significant longer term thermal events were about 1000 years ago, 10000 years ago, and 11 thousand years ago [that whole Venus incident]. This brings us to the second thing to look at. There are thermal spikes that last for 100 years or less. Although short lived, the thermal change during these times is as much as 6 degrees centigrade. That factor is significant when viewing a later chart.

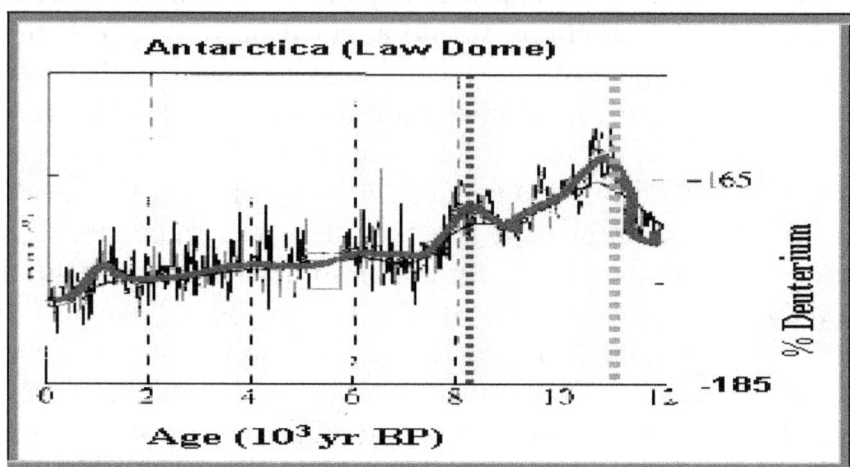

The second chart, next, shows the temperature of Antarctica over 5 different sites. By cross comparing the data, we can be assured that the data has a high degree of accuracy.

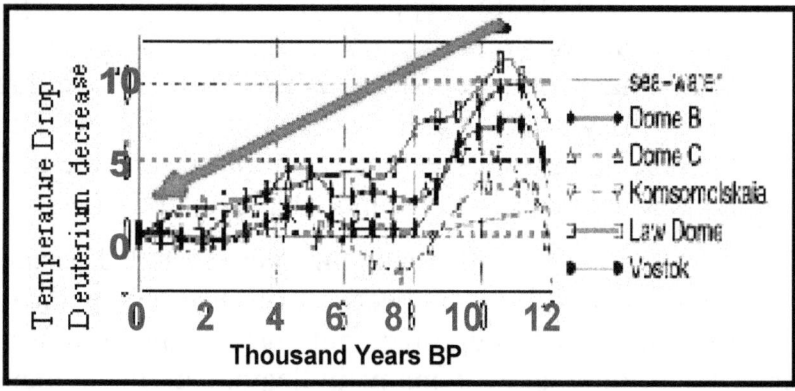

Instead of the long term ice core graph being shown to us, the short term one below is typically provided to make us think the opposite is happening.

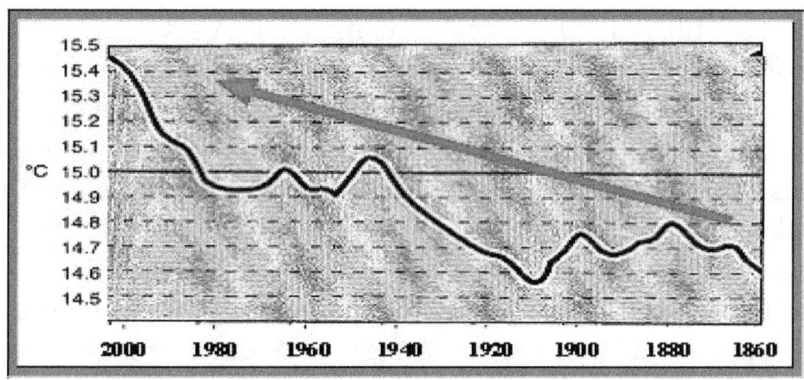

The graph shows a rise in temperature at a specific area of about 1 degree in 150 years, of course the accuracy of the measurements is most likely not much better than ½ degree, so the almost flat slope may be even flatter and it should be going the other way in a few years even if we don't find a way to keep the cattle methane flatulence from escaping into the atmosphere. This seems to represents one of the thousands of thermal spikes indicated in the Ice Core samples—nothing more.

## Dinosaur Flatulence

As an aside, the federal government is spending multiple hundreds of thousands of dollars researching one of the largest producers of ozone destroying methane—the cow butt. This whole concept of Cattle flatulence destroying our ozone got me to thinking about dinosaurs again. As much as cattle expel methane, dinosaurs must have been awful to travel behind.

From the high levels of the almost impossibly fossilized remains and the huge quantities of oil left behind we can be assured that there were more dinosaurs on this planet than we could possibly imagine. From them, the methane levels in the very ancient past surely would have been enough to de-ozone-ate the atmosphere and the old earth should have been consumed in a ball of fire like many scientist indicate we are heading for if we don't stop up the pesky cattle butts. I think we should leave cows alone. It didn't happen during the ancient smelly time and the likelihood of it happening today is slim to none. We are probably going the other way.

## Destroying Trees

Scientists still don't know what causes Ice Ages much less can they devise ways to slow its progress. We know that the bogus warning for people to halt cutting trees or the earth will get too warm didn't halt tree cutting. The idea of having too much $CO_2$ because there are not enough trees to consume it is a bad thing, but like the cattle methane, it should not be brought out as a reason for the earth getting warmer. The earth is still doing what it wants to do. One thing that does absolutely cause the homeostasis of the earth to change is a change in its axis of rotation. If a change like that occurs, plants won't grow as much because the temperature changes and other environmental changes disrupt growing cycles, less coverage of the earth means the earth temperature will drop in temperature significantly rather than increase and large masses of ice would be repositioned from one location to another which would further amplify a delicate situation. This may be the way an Ice Age normally begins. By the way I'm all for halting the extermination of our rain forest, because I like the medicines we get from them, and the oxygen they produce. Fewer trees is not good for any of us.

# What Can We Expect Next

The answer is not what you would like to hear it begins with another shift in our poles and predictions of a comet or large mass hitting the earth. Huge wars are predicted and almost total destruction. Pretty nasty. While this book is not going into those elements directly, let's look at the evidence of our impending polar shift as it is an indicator of things to come.

## Pole Shift Catastrophe

Quite a few researchers are now indicating that our next destruction period on earth will not come from the sky, nor will it come from war or even a simple Ice Age. Our next global catastrophe may be another one of those shifts in the rotational axis that caused the Mammoths to freeze a few thousand years ago. These scientists have cause to instigate concern as we look at the findings below that are not widely disseminated.

## Magnetic Pole Wander

As I mentioned before, the magnetic pole has wandered all over the place, and you may think that the wandering was only in ancient times, but we know how dramatic this movement is because of the work of Paul Serson and Jack Clark, of the Dominion Observatory. The pole wanders daily in a roughly elliptical path around its average position, and may frequently have movements as much as 80 km away from this position when the Earth's magnetic field is disturbed. Accurate observations by Canadian government scientists in 1962, 1973, 1984, and most recently in 1994, showed that a northwesterly motion of the pole is continuing, and that during this century it has moved on average 10 km per year. The chart below is a running 30 day plot of the readings taken from one of several fluxgate magnetometer sensing sites placed around Canada and Alaska to check for movement of the earth's magnetic field.

- The component on the top left is positive magnetic northward

- The component on the bottom left is positive eastward

Note the wild movement of the Earth's magnetic field as described by this instrument. First the field jumps to the Northwest then moves back eastward, followed by a southward travel and then switching east.

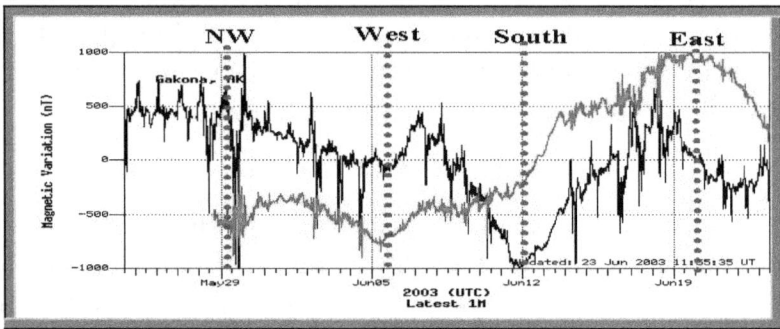

Our Polar Satellite data confirms magnetic movement. The graph below shows magnetic wander about the North Pole as captured from space. The general trend towards the northwest can easily be seen.

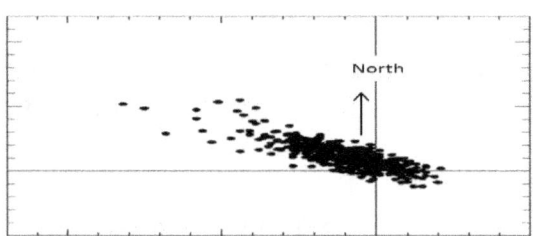

**Scary Fluctuations**

OK! Here comes the scary part. Since 1994 the average speed of the magnetic drift has increased to an average of 15 km per year which makes me sit up a little, but the story gets even more disturbing for those of us who would like to stay in the same climatic position. Satellite data has now been used to compare the strength and direction of the magnetic field in 1980 and again in 2000. According to researchers at "Physics of the Globe Institute" of Paris and the "Danish Space Research Institute", the magnetic field off the southern tip of Africa has already moved dramatically. I know that needs a little explaining. For that we go to Gauthier Hulot, a member of the Danish research team. He and his teammates indicated that molten iron under Africa is now moving in a direction, which will gradually weaken the dominant magnetic field and then reverse it. If the trend continues, the research shows that, we could be seeing the first steps toward a new North Pole. The timing has not been brought out, but saying "any time in the near future" is not good for me. I'm quite used to the hot Florida Weather where I live.

Researchers have determined that the ultimate cause of the magnetic fluctuations is the Sun. The Sun constantly emits charged particles that, on encountering the Earth's magnetic field, cause electric currents to be produced in the upper atmosphere. These electric currents disturb the magnetic field, resulting in a temporary shift in the pole's position. The distance and speed of these displacements will, of course, depend on the nature of the disturbances in the magnetic field, but they are occurring constantly. With the earth being generally a sphere and most of the "Iron" component being in a liquid state, the magnetic element of the earth has no particular position it likes or doesn't like. Therefore, it could shift at any time. Whenever the magnetic pole shifts, the rotation axis shift will soon follow as the spinning creates a differential voltage that can only be eliminated by having both elements in the same plane. The shift does not have to be very significant for a destabilizing effect and "Ice Age" to occur so get out your long underwear.

### Gloom and Doom

I know it all seems like gloom and doom, but all I'm trying to say here is that history repeats itself and the signs indicate that some of the repeating is coming fairly soon. The Earth has gone through many Meteor attacks, Axis-of-spin shifts, close encounters with other planets, and dinosaur flatulence. Except for the dinosaur thing, more of each are coming again and putting our head in a hole won't stop it and won't delay it. There may be ways to prepare for some of the disasters, but certainly not if we don't know about them. Take you head out of the sand and make sure that you get the creator's ear, because the signs indicate that time is short. Travelling to other worlds is not a great option, because they fared worse than earth during the ancient episodes. The best thing is to do is understand what happened in the past and realize that it WILL happen again.

## Revelation Gloom

The book of Revelation in the Bible says that very soon heaven will be destroyed and so will the earth. It gives us hope by saying that there will be a new heaven and a new earth to come. I'm sure the new one will go through the same turmoil and the planets will also reemerge but that is another story.

# Conclusion

The model of a massive explosion bringing forth the galaxies and initiating our Solar system for the most part works. We had to add in a second universe and some dimensions that we can't see to get it into workable condition, but the model seems workable. While I didn't go into details of what might have happened in the "other universe" during the development of ours, there is a substantial amount of ancient literature devoted to wars between an unseen heaven and the world to fill our imaginations.

## Not Seeded From a Distant Galaxy

This book concentrated on the development or evolution of the planets and the planets that may have supported life in particular. The idea of our solar system being "seeded" by some inhabitants of a faraway galaxy makes no since to me. There are too many "whys" and "hows" to make it reasonable. If the seeding was over millions of years, why come back to the same planet group? If the action was done to colonize the universe for protection, where's the protection? If the problem was planetary overcrowding, then some level of communication would still be in place. However, there may have been some attempts in the past.

## We are not Fish-men

I know that the Dogan tribe of Mali have explicit details of fish-men coming to earth from a planet going around Sirius and that the Sumerian belief in a half fish half man caretaker sounds fishy. These beings might have tried to colonize earth, but humans survived and they did not. By the way, do not discount the punishment of the "serpent" guy in the Garden of Eden who was turned into a reptile. Around the world researchers have found statue after statue of half reptilian-half human people holding their children. Possibly, the fish-men were really some of the descendants of the reptilian punished group of people and they were not from the stars.

## We are not Reptile People

We know that these Reptilian-like humans lived in Mohen jo Daro, and in parts of Iran, and Iraq during ancient times and probably many other places. We also know that this particular group procreated on earth and had become part of several communities long ago. Archeologists have found hundreds of statuary pieces and written testimony from multiple societies to confirm the existence, but we are not reptile beings. They could not adapt in some way and humans again were the sole users of earth.

These were not superior beings but similar. Humans were the advanced race in the ancient times. There is too much evidence that shows humans were much more advanced than we are today and that this civilization was as much as 300 thousand years ago. While I did not present the vast amount of data confirming this fact, let me assure you that the only way science has addressed most of the details is to simply say the evidence is an anomaly.

## The Planets Are Not Stable

Mars and Earth messed each other up and then earth and Venus had their time. I know all this seems fanciful, but to come up with plate tectonic mountain making and greenhouse effect planet incinerating is not an honest way to deal will problems.

## Flying Man

The ancient humans had a rough time on earth at times. Eleven major extinction periods alone would have set back the advancement of society unless the leaders and others left during the really bad times. While I did not go over the huge amount of evidence about man's ability to fly and the many designs of flying vessels that were depicted during ancient times, most people know about Ezekiel's "wheel in a wheel" vehicle that was recorded in the Biblical history. Just be aware that there is an enormous amount of data about man's flying.

## Planetary Colonization

The evidence indicates that man flew to the surrounding planets and colonized many of them. The apparent building shapes are man-like buildings. The Pyramid shapes are man-like pyramid shapes. The cities are man-like cities with man-like streets. The requirement for man-needed water seems to be present at locations of colonization. In areas where air had to be manufactured for man-like existence, there seems to be underground or sealed environments.

As with all man's accomplishments, war destroyed many. Over and over major wars destroyed one group after another. The evidence of nuclear inclusion is very apparent in ancient literature and physical evidence. This means that there would have been no winners.

## Babel Story Revisited

The Biblical Tower of Babel story is interesting as we think about losses from war. Just prior to the tower thing, this mighty man Nimrod essentially took over the Middle East and began to build a tower in Lebanon [I know some of you think it was in Babylon, but the evidence does not support it] Why would God have destroyed so many people for building a tower? Would he have been afraid of a simple building? The indication is that the now earthbound humans were again beginning to wage war in space and the launching platform tower was scuttled to eliminate more devastation. The destruction  on earth of the incident is interesting as well. Biblical and ancient Jewish testimony indicates that thousands were immediately destroyed, but 1/3 were changed into something like monkeys while everyone else had lost the ability to communicate except to close family members. To me this sounds like humans had ALL lost major capabilities of their brains during this ordeal some lost more than others. The evidence suggests that we have never regained those capabilities whatever they were. Today scientists tell us we can only use 1/10th of our brains for some odd reason. Don't you think that is odd?

## Back to the Planets

The evidence tells us that colonists on several of the planets came from earth, traded with earth, fought with earth, and were destroyed by various things. Some were destroyed by war and massive blast streaks were left behind. Others succumbed to the extinction by planetary instability, loss of atmosphere and many other dilemmas. The question might be, "Is there any life on the planets today?"

## The Old UFO Story

The recent incidents of UFO sightings, cattle mutilations, visitations and all the other stuff, might be an indication that some HUMANS still are surviving in some way. These visitors could not come from far away galaxies. The frequency of the events doesn't make sense. The dissection of cattle also provides us with an indication that the VISITORS are human. Cattle DNA are similar to human. In fact, scientist are looking for ways to change cattle blood for human use and other oddball things showing similarity. The human visitors might experiment on nonhuman subjects to gain insight but would not want to indiscriminately kill other humans. If these were more of the reptile beings from long ago, we would see dissected humans as well, I believe.

We might not be alone, but our friends are, most likely, human. During the upcoming catastrophes that will surely come, I hope that we will get support from these ancestors.

The End

# About the Author

Steve Preston is a long lime author of scientific, esoteric facts. His series on the creation of mankind is shown below. The series focuses on the painful truths rather than whitewashed details that make us comfortable. If you are interested in the truth instead of comfort, please continue to read and, while you are at it, review other works by Mr. Preston as shown below.

**Four Part Series "Vibrational Matter"**
*Vibrational Matter*
*10-Dimentional Universe*
*Walk Through a Wall and Time*
*Meaning of Light, Life,& Death*
*Live and Die the Right Way [Addendum]*

**Eight Part Series "History of Mankind"**
*The First Creation of Man*
*The Second Creation of Man*
*The Creation Of Adam And Eve*
*The Antediluvian War Years*
*Man After the Flood*
*Life After the Babel War*
*A New View Of Modern History*
*The 20th Century To The End Of Time*

**Truth Series**
*The Truth About Dinosaurs*
*The Truth About The Earth*
*7 Destructions of the Earth*
*Truth About the Heaven War*
*Truth About Dinosaurs*
*Who Really Discovered the Americas?*
*God Didn't Make The Ape*

*Our Very Odd Presidents*
*Today's Monsters*
*Truth About Vampires*
*Living Underground*

## Less Ancient Works

*A Closer Look At Lincoln*
*Adam, Lilith, and Eve*
*America's Civil War Lie*
*Ancient History of Flying*
*Behind the Tower of Babel*
*The Funny Book of Law*
*When Giants Ruled the Earth*
*Lizard People*

## Planet and Odd Series

*When Did People Live on The Moon?*
*Evolution of the Planets*
*The Day Venus Exploded*
*Living on Mars*
*The Book Of Odd*
*More Oddness*
*Why Are There So Many Anomalies?*
*Stupid Science*

## Religious Series

*Self, Soul and Spirit*
*Truth About the Anakim Gods*
*A Closer Look At Genesis*
*Genesis Companion*
*History of the World Confirmed by the Bible*
*Bible Inspiration Yes or No*
*Kingdoms Before the Flood*
*Four Armegeddons*

www.ingramcontent.com/pod-product-compliance
Lightning Source LLC
Chambersburg PA
CBHW051904170526
45168CB00001B/235